CONTEMPORARY MATHEMATICS

142

Several Complex Variables in China

Chung-Chun Yang
Sheng Gong
Editors

American Mathematical Society
Providence, Rhode Island

Math
Sep/ae

1991 *Mathematics Subject Classification*. Primary 32-02, 32Axx, 32Cxx, 32Hxx, 32Mxx.

Library of Congress Cataloging-in-Publication Data

Several complex variables in China/[edited by] Chung-Chun Yang, Sheng Gong.
p. cm.–(Contemporary mathematics, ISSN 0271-4132; v. 142)
ISBN 0-8218-5164-0
1. Functions of several complex variables. 2. Mathematics—Research—China. I. Yang, Chung-Chun, 1942– . II. Gong, Sheng, 1930– . III. Series: Contemporary mathematics (American Mathematical Society); v. 142.
QA331.7.S47 1993
515′.94–dc20

92-44828
CIP

Sd 10/6/93 ES

Recent Titles in This Series

(Continued in the back of this publication)

Contents

Preface

The study of the theory of functions of several complex variables in China was founded and largely developed by the research of the late Prefessor Hua Lo-Keng. Since the early 1940's, mathematicians have increasingly been making connections among different branches of mathematics. Frequently the results have been a more difficult and more abstract theory. When these developments evolved, Hua had already started important research on the holomorphic automorphic functions of the classical domains of several complex variables. After spending several years of teaching and doing research in the United States, Hua returned to China in the early 1950's and vigorously promoted research in the area of several complex variables. He organized seminars on the topic at the Institute of Mathematics, Academia Sinica. These seminars initiated and promoted the study of several complex variables in China. Even today the impact of his efforts are easily observed. Hua's theories and techniques are prevalent among contemporary research in several complex variables in China and abroad. The survival of Hua's contributions is particularly remarkable since mathematics, as other scientific disciplines, was interrupted by the eruption of the ten-year Cultural Revolution of China which ended in 1976. Since then, China has adopted a more open foreign policy and has promoted academic exchanges with other countries. As a result of these developments, modern views of the theory of several complex variables and the associated results and methods have been introduced into China. The well founded group interested in the theory of several complex variables quickly regained its momentum and entered into the new era in the study equipped with the newly gained knowledge. Consequently, there were many important new results in the theory of several complex variables obtained in China. However, until the latest decade, most of the mathematical research findings were written in Chinese and announced only in domestic journals and books. This limited communication of results prevented the Western World from learning and using the new theory of several complex variables developed by the Chinese. To announce many of these lesser known important results is one of the main reasons for publishing this collection.

In 1935, E. Cartan proved there are precisely six types of irreducible homogeneous bounded symmetric domains; two of these are exceptional, which only occur in dimensions 16 and 27. Hua called the other four types of domains "the classical domains." These domains consist of matrices (with complex entries) which satisfy certain conditions. In the first three cases the conditions are given in terms of the positive definiteness of certain matrices.

(1) $\mathcal{R}_I : I - Z\bar{Z}' > 0$, where Z is an $m \times n$ matrix;

(2) $\mathcal{R}_{II} : I - Z\bar{Z} > 0$, where Z is an $n \times n$ symmetric matrix;

(3) $\mathcal{R}_{III} : I + Z\bar{Z} > 0$, where Z is a skew symmetric matrix of order n;

(4) $\mathcal{R}_{IV} : |ZZ'|^2 + 1 - 2\bar{Z}Z' > 0$, $|ZZ'| < 1$, where Z is a $1 \times n$ matrix with $n \geq 2$.

Around 1943, Siegel and Hua, independently studied the automorphic functions of the classical domains and obtained many important results; some of which overlapped. However, Hua's approaches and results were typified by his direct style and concrete formulations, and by difficult but ingenious computations. Hua's well known monograph, *Harmonic Analysis of Functions of Several Complex Variables in the Classical Domains*, was written in Chinese during 1955. It was immediately translated into Russian and English, and became a classical work. In that book, Hua, skillfully used analysis and matrix methods. Using group representation theory as a tool, he derived concretely complete orthonormal systems, groups of holomorphic automorphisms, Bergman kernels, Cauchy kernels, Poisson kernels, and so on for the classical domains. As was pointed out in W. Rudin's book, *Function Theory in the Unit Ball of $\mathbb{C}^n$*, even fundamental results such as the explicit Cauchy Integral Formula on the unit ball of $\mathbb{C}^n$ were contributions of Hua. Moreover, Hua's book greatly influenced developments in the theory of representations of Lie groups, the theory of homogeneous spaces, and the theory of automorphic functions of several complex variables. It is also noteworthy that the technical machinary which Hua introduced in his book included several classes of algebraic identities and the computations of integrals of functions of a matrix argument. Each these topics is of independent interest and has been studied by many research mathematicians throughout the world.

Besides the book of Hua just discussed, another of Hua's books has contributed importantly to the development of mathematics in China. This book is *Starting on the Unit Circle*. It was first written in Chinese in 1977; an English translation was published by Springer-Verlag in 1981. Inspired and fortified by Hua's outstanding achievements, the study of several complex variables on the classical domains and related topics have become major and active fields of mathematical research in China. Hua's followers have made many significant contributions, especially in the study of functions in symmetric domains which inlcude topics in the theory of harmonic functions, generalizations of Schwarz' lemma, the theory of singular integrals and of singular integral equations, the study of integral representation for an arbitrary domain and its boundary behavior, harmonic analysis on classical groups which are characteristic manifolds for classical domains and harmonic analysis on compact Lie groups, the representation of the two exceptional irreducible bounded symmetric domains and its Bergman, Cauchy, and Poisson kernels, the classification of certain Siegel domains (as generalizations of the classical domains), the study of geometric theory of biholomorphic mappings, the systems of two partial differential equations of two functions with two independent variables, etc.

The study of complex geometry in China was initiated by two discoveries of Hua in the 1950's. He proved that the Riemannian curvature in the Bergman metric of a bounded domain is at most two and that the holomorphic domain of constant curvature can be mapped biholomorphically onto the unit ball. After that, Chinese mathematicians did lots of interesting works on complex geometry. Currently complex geometry continues to be very popular and to develop rapidly in China.

Since Hua's work is well known, our intention will be to present to Western readers more recent works of Chinese mathematicians. We regret that for various reasons some work of Chinese researchers may have failed to be included in this collection. However, we do believe that the articles presented here will introduce the reader to some unique features of the study of several complex variables in China, including some results obtained recently by oversea Chinese mathematicians. Also these articles should give the reader some appreciation of the dynamic growth of this theory in China. Thus, it is hoped, closer and more beneficial contacts between Eastern and Western researchers will result in the advancement of knowledge in the field of several complex variables.

The editors would like to express their sincere thanks to Dr. Weiqi Gao for the time he spend in helping us edit this volume.

S. Gong 龚升

C. C. Yang 杨重骏

Contributors

Zhi-Hua Chen　陈志华
Department of Applied Mathematics, Shanghai Jiaotong University, Shanghai 200030, People's Republic of China

Sheng Gong　龚升
Department of Mathematics, University of Science and Technology of China, Hefei, Anhui 230026, People's Republic of China

Shanyu Ji　嵇善瑜
Department of Mathematics, University of Houston, Houston, TX 77204, U. S. A.

Bao Qin Li　李宝勤
Department of Mathematics, University of Maryland, College Park, MD 20742 U. S. A.

Min Ru　汝敏
Department of Mathematics, National University of Singapore, Kent Ridge Crescent, Singapore 0511, Republic of Singapore

Ji-Huai Shi　史济怀
Department of Mathematics, University of Science and Technology of China, Hefei, Anhui 230026, People's Republic of China

Yichao Xu　许以超
Institute of Mathematics, Academia Sinica, Beijing 100080, People's Republic of China

Chung-Chun Yang　杨重骏
Department of Mathematics, The Hong Kong University of Science and Technology, Clear Water Bay, Kowloon, Hong Kong

Zhimin Yan 严志敏

Department of Mathematics, University of California at Berkeley, Berkeley, CA 94720, U. S. A.

Jinhao Zhang 张锦豪

Department of Mathematics, Fudan University, Shanghai 200433, People's Republic of China

Tongde Zhong 钟同德

Department of Mathematics, Xiamen University, Xiamen, Fujian 361005, People's Republic of China

Contemporary Mathematics
Volume **142**, 1993

Complex Geometry in China

ZHI-HUA CHEN

§1. The classical Schwarz lemma, first published by Carathéodory [1] in 1912, states that a holomorphic function $f(z)$ of one complex variable with $f(0) = 0$ and $|f(z)| \leqslant 1$ in the unit disc, satisfies $|f(z)| \leqslant |z|$ and $|f'(0)| \leqslant 1$. After four years, G. Pick [2] reformulated the Schwarz lemma to show that every holomorphic map of the unit disc into itself is distance decreasing in the Poincaré metric. We recall that the Poincaré metric for the unit disc is

$$ds^2 = \frac{|dz|^2}{(1 - |z|^2)^2},$$

and the unit disc is complete with respect to this metric. G. Pick's version of the Schwarz lemma reads

$$f^* ds^2 \leqslant ds^2,$$

$\forall f \in \operatorname{Hol}(\Delta, \Delta)$, where Δ denotes the unit disc and $\operatorname{Hol}(X, Y)$ denotes the set of all holomorphic maps from X to Y.

According to Pick's idea, Lu [3] established the corresponding Schwarz lemma for four kinds of classical domains $\mathcal{R}_I$, $\mathcal{R}_{II}$, $\mathcal{R}_{III}$ and $\mathcal{R}_{IV}$, where

$$\mathcal{R}_I = \left\{ Z \in \mathbb{C}^{m \times n} \mid I^{(m)} - Z\bar{Z}' > 0, m \leqslant n \right\},$$

in which $\mathbb{C}^{m \times n}$ denotes the space of the $m \times n$ complex matrix,

$$\mathcal{R}_{II} = \left\{ Z \in \mathbb{C}^{p \times p} \mid I^{(p)} - Z\bar{Z}' > 0, Z = Z' \right\},$$

$$\mathcal{R}_{III} = \left\{ Z \in \mathbb{C}^{q \times q} \mid I^{(q)} - Z\bar{Z}' > 0, Z = -Z' \right\},$$

and

$$\mathcal{R}_{IV} = \left\{ Z \in \mathbb{C}^{1 \times n} \mid 1 + |ZZ'|^2 - 2Z\bar{Z}' > 0, 1 - |ZZ'| > 0 \right\}.$$

In fact, Lu proved that if $D \subset \mathbb{C}^n$ is a bounded transitive domain which contains the origin point 0 and ds^2 is its Bergman metric then there exists a Schwarz

1991 *Mathematics Subject Classification.* 32C17, 32C10, 32-02.
Research supported by the National Science Foundation of China.
This paper is in final form and no version of it will be submitted for publication elsewhere.

constant $k_0(D) > 0$ which depends only on D such that $\forall f \in \mathrm{Hol}(D, D)$ with $f(0) = 0$, $f^* ds^2 \leqslant k_0(D) ds^2$. The Bergman metric of any classical domain is invariant under the holomorphic automorphism group of the domain. Using this observation, Lu [4] obtained the values of the corresponding Schwarz constants of the four classical domains, i.e. $k_0(\mathcal{R}_I(m, n)) = \sqrt{m}$, $k_0(\mathcal{R}_{II}(p)) = \sqrt{p}$, $k_0(\mathcal{R}_{III}(q)) = \sqrt{[q/2]}$ and $k_0(\mathcal{R}_{IV}(n)) = \sqrt{2}$.

§2. The connection between curvature and the metric decreasing property expressed in the Schwarz-Pick lemma was never discovered until L. Ahlfors [5]. He established the following theorem: If $d\sigma^2$ is a differential metric on Δ, whose Gauss curvature is bounded above by -4, then $\forall f \in \mathrm{Hol}(\Delta, \Delta)$

$$f^* d\sigma^2 \leqslant ds^2,$$

where ds^2 is the standard Poincaré metric on Δ. This theorem show that the metric decreasing property or the decreasing coefficient (Schwarz constant) of the holomorphic maps depends on the curvatures of the origin manifold and the target manifold.

Yau [6] was the first one to establish the Schwarz lemma on complex manifolds. Following the idea of L. Ahlfors, Yau proved:

Suppose M is an m-dimensional complete Kähler manifold, whose Ricci curvature is bounded from below by a constant R_1 and N is an n-dimensional Hermitian manifold, whose biholomorphic sectional curvature is bounded from above by a negative constant C_2. Moreover, if there exists a nontrivial holomorphic map f from M into N, then $R_1 < 0$ and

$$f^* ds_N^2 \leqslant \frac{R_1}{C_2} ds_M^2.$$

In [6], Yau also proved that if $ds_M^2 = \theta^i \otimes \bar{\theta}^i$, $ds_N^2 = \omega^\alpha \otimes \bar{\omega}^\alpha$, $f^* \omega^\alpha = a_i^\alpha \theta^i$ and $m \times n$ matrix $A = (a_i^\alpha)$ is the representation of df respect to the unitary frame, then $u = \mathrm{tr}(A\bar{A}')$ is a differentiable function on M. Yau [6] used Omori's generalization of the maximal principle to u, an upper bound of u was obtained and

$$f^* ds_N^2 \leqslant u \, ds_M^2 \leqslant \frac{R}{C_1} ds_M^2.$$

Various higher-dimensional generalizations of the Schwarz-Pick-Ahlfors type of results were obtained by Chen-Cheng-Lu in [7]. They considered a complete Kähler manifold M whose holomorphic sectional curvature is bounded from below by a non-positive constant K_1 and whose sectional curvature is also bounded from below and Hermitian manifold N whose holomorphic section curvature is bounded from above by a negative constant K_2. Then $\forall f \in \mathrm{Hol}(M, N)$

$$f^* ds_N^2 \leqslant \frac{K_1}{K_2} ds_M^2.$$

In the above theorem, the holomorphic sectional curvature assumptions were used instead of the Ricci curvature assumption on M and the biholomorphic sectional curvature assumption on N in [6]. Moreover the Schwarz constant of the theorem is sharp. For example, when M and N are the same classical domains, then Theorem 16 in [8] showed $\dfrac{K_1}{K_2}$ is the best bound there.

§3. In [7], a local version of Schwarz lemma was first established. Let $B_a(z) \subset M$ be a geodesic ball with center z and radius a, $\lambda(x)$ be the largest eigenvalue of the matrix $A(x)\overline{A(x)}'$, where $A(x)$ is the matrix relative to the holomorphic map $f : M \to N$ as cited above. Then consider the continuous function $\Phi(x) = (a^2 - \gamma^2(x))^2 \lambda(x)$ on $B_a(z)$, where $\gamma(x)$ is the distance between x and z. It is trivial to show that $\Phi(x)$ attains its maximum at some point $x_0 \in B_a(z)$. Now, $\xi(x_0)$ is a unit eigenvector corresponding to $\lambda(x_0)$, and ξ is a vector field defined near x_0 such that $\xi(x)$ is parallel to $\xi(x_0)$ as x moves along the geodesic from x_0 to x. Now, define

$$\tilde{\lambda}(x) = |A\xi|^2.$$

Then $\tilde{\lambda}(x)$ is smooth near x_0 and such that $\tilde{\lambda}(x_0) = \lambda(x_0)$ and $\tilde{\lambda}(x) \leqslant \lambda(x)$. Hence $(a^2 - \gamma(x)^2)^2 \tilde{\lambda}(x)$ also attains its maximum at x_0. Thus the maximum condition can be applied to $(a^2 - \gamma^2)^2 \tilde{\lambda}(x)$ at x_0 to derive the Schwarz lemma on $B_a(x)$, as $a \to +\infty$.

Yang and Chen [9] gave another type of Schwarz theorem: Let M be a complete Kähler manifold, whose Ricci curvature is bounded below by a non-positive constant R and N be a Hermitian manifold with non-positive holomorphic bisectional curvature and whose holomorphic sectional curvature is bounded from above by a negative constant K. Then $\forall f \in \mathrm{Hol}(M, N)$

$$f^* ds_N^2 \leqslant \frac{R}{K} ds_M^2.$$

This theorem was derived as a consequence of a refinement of the original argument of Yau in [6]. Note also where A is the matrix df relative to the unitary frame Yau used the generalized maximum principle of H. Omori on the function $u = \mathrm{tr}(A\bar{A}')$. Yang-Chen considered the more complicated function $u = [\mathrm{tr}(A\bar{A}')^k]^{1/k}$, where k is any large integer; especially, they considered its behavior when $k \to +\infty$.

§4. Methods similar to the above approach have been used to investigate various kinds of harmonic maps. For instance, in [9] and [10], the following theorems A and B concerning harmonic maps were obtained.

THEOREM A. *Suppose $f : M \to N$ is k order q-quasi conformal harmonic map from M into N, where M is a complete Riemannian manifold with Ricci curvature bounded from below by R, and N is a Riemannian manifold with the sectional curvature bounded from above by $K < 0$. Then*

$$f^* ds_N^2 \leqslant \frac{q^2}{k-1} \frac{R}{K} ds_M^2.$$

The original estimate in [11] *is* $f^* ds_N^2 \leqslant \dfrac{kq^4}{k-1} \dfrac{R}{K} ds_M^2$.

THEOREM B. *Suppose M is a complete Riemannian manifold with Ricci curvature bounded from below by a non-positive constant $-A$ $(A \geqslant 0)$ and N is a Riemannian manifold with sectional curvature bounded from above by a negative constant $-B$ $(B > 0)$. If $f : M \to N$ is a K-dilation harmonic map from M into N, then $f^* ds_N^2 \leqslant K^2 \dfrac{A}{B} ds_M^2$ and $f^* V_N \leqslant \left(\dfrac{(1+K^2)A}{mB} \right)^{m/2} \cdot V_M$.*

Note that Goldberg, S. I. and Har'EL, Z. [12] obtained $f^* ds_N^2 \leqslant p^2 \dfrac{K^2 A}{2B} ds_M^2$, where p is the rank of f and $f^* V_N \leqslant \left(\dfrac{mK^2 A r}{2B} \right)^{m/2} V_M$, respectively.

§5. The Kähler condition on M has been a common assumption in the proofs of the Schwarz lemmas mentioned above. Yang and Chen [13] considered the origin manifold M to be a Hermitian manifold. In this case, let

$$ds_M^2 = \theta^i \otimes \bar{\theta}^i$$

and structure equations with respect to this unitary coframe $\{\theta^i\}$ be

$$d\theta^i = \theta^i \wedge \theta_j^i + \frac{1}{2} T_{jk}^i \theta^j \wedge \theta^k,$$

$$d\theta_i^j - \theta_i^k \wedge \theta_k^j = \frac{1}{2} R_{ik\ell}^j \theta^k \wedge \bar{\theta}^\ell,$$

$$T_{jk}^i = -T_{ik}^j = T_{j\bar{i}k},$$

$$R_{ik\ell}^j = R_{i\bar{j}k\bar{\ell}}, \quad \overline{R_{i\bar{j}k\bar{\ell}}} = R_{j\bar{i}\ell\bar{k}}.$$

Before stating further results, let us introduce some notations. Set $R_{i\bar{j}} := R_{k\bar{k}i\bar{j}}$, $R_{i\bar{j}}^T := R_{i\bar{j}k\bar{k}}$ and for unit $(1,0)$ vectors $\xi, \eta \in T_x^{1,0}(M)$, set

$$R(\xi, x) = R_{i\bar{j}} \xi^i \bar{\xi}^j, \quad R^T(\xi, x) = R_{i\bar{j}}^T \xi^i \bar{\xi}^j,$$

and $T_k(\xi, \eta; x_0) = T_{i\bar{j}k} \xi^i \bar{\eta}^j$, $\|T(\xi, \eta; x_0)\| = \left(\sum_k |T_k(\xi, \eta, x_0)|^2 \right)^{1/2}$. Let $B_a(z)$ be a geodesic ball with center z and radius a, and

$$K_a := \inf_{\substack{x \in B_a(z) \\ \xi, \eta \in T_x^{1,0}(M)}} K(\xi, \eta; x),$$

$$R_a^T := \inf_{\substack{x \in B_a(z) \\ \xi \in T_x^{1,0}(M)}} R^T(\xi, x),$$

$$T_a := \sup_{\substack{x \in B_a(z) \\ \xi, \eta \in T_x^{1,0}(M)}} \|T(\xi, \eta; x)\|$$

where $K(\xi, \eta; x)$ is the holomorphic bisectional curvature spanned by $\xi \wedge \eta$, $R^T(\xi, x)$ is called the second Ricci curvature decided by ξ.

Then using the method similar to the one used in [7] the following Theorems C, D, and E were obtained.

THEOREM C. *Let M be a complete Hermitian manifold whose holomorphic sectional curvature is bounded from below by K_1 with holomorphic bisectional curvature and torsion satisfying*

$$\lim_{a \to \infty} \frac{K_a}{a^2} = 0, \ and \ \lim_{a \to \infty} \frac{T_a}{a} = 0$$

respectively. Let N be a Hermitian manifold whose holomorphic sectional curvature is bounded from above by $K_2 < 0$, then $\forall f \in \mathrm{Hol}(M, N)$

$$f^* ds_N^2 \leqslant \frac{K_1}{K_2} ds_M^2.$$

THEOREM D. *Let M be a complete Hermitian manifold whose second Ricci Curvature is bounded from below by R_1^T with torsion satisfying $\lim_{a \to \infty} \frac{T_a}{a} = 0$. Let N be a Hermitian manifold whose holomorphic sectional curvature is bounded from above by $K_2 < 0$ with non-positive holomorphic bisectional curvature. Then $\forall f \in \mathrm{Hol}(M, N)$*

$$f^* ds_N^2 \leqslant \frac{R_1^T}{K_2} ds_M^2.$$

THEOREM E. *Let M be a complete m-dimensional Hermitian manifold whose scalar curvature is bounded from below by S_1 with its second Ricci curvature and torsion satisfying respectively,*

$$\lim_{a \to \infty} \frac{R_a^T}{a^2} = 0, \ and \ \lim_{a \to \infty} \frac{T_a}{a} = 0.$$

Let N be an m-dimensional Hermitian manifold whose Ricci curvature is bounded from above by $R_2 < 0$, then $\forall f \in \mathrm{Hol}(M, N)$,

$$f^* dV_N \leqslant \left(\frac{S_1}{mR_2} \right)^m dV_M.$$

§6. From the above analysis, it is easy to realize that when the various curvature assumptions of the original manifolds have a non-negative lower bound, then $f^* ds_N^2 \equiv 0$, i.e. f must be a constant map. This is the simplest Liouville type theorem. In fact, there are several versions of Liouville type theorems. For instance, for every complete Hermitian manifold N, whose sectional curvature $\leqslant \frac{1}{4\rho(z)^2}$, where ρ is the distance between z and given point 0, and for each geodesic ball $B_a(0)$ with center 0 and radius a, $a < +\infty$, one can endow on $B_a(z)$ a new metric such that the corresponding curvature is bounded above by a negative constant. Therefore if the curvature on M is non-negative, $f : M \to N$ is holomorphic map and $f(M) \subset\subset N$, then f must be a constant.

According to this observation, Chen and Yang [14] obtained several Liouville type theorems on complex manifolds. We list four such results as follows.

THEOREM. *Let M be a complete Hermitian manifold with a non-negative unitary curvature and bounded torsion, such that its holomorphic bisectional curvature is bounded from below. Let N be a Hermitian manifold with its sectional curvature $\leqslant \dfrac{1}{4p^2}$ and self-torsion $< \dfrac{1-2\varepsilon}{2p}$. If $f \in \mathrm{Hol}(M,N)$ such that $f(M) \subset B_a(z)$, then f must be constant.*

THEOREM. *Let M be a complete Hermitian manifold with non-negative Ricci curvature and bounded torsion, and let N be Hermitian manifold satisfying the conditions cited in the above theorem. Then there exists no non-constant bounded holomorphic maps from M to N.*

THEOREM. *Let M be a complete Hermitian manifold with bounded torsion and which satisfies one of the following conditions:*

   (1) *M has non-negative unitary curvature and its holomorphic bisectional curvature is bounded from below.*
   (2) *M has non-negative second Ricci curvature.*
   (3) *M has non-positive scalar curvature.*

Let N be a complete Hermitian manifold with its sectional curvature and self-torsion $< \dfrac{1-\varepsilon}{2\rho}$, for some $\varepsilon > 0$. Then there exists no non-constant bounded holomorphic maps from M to N.

THEOREM. *Let M be a complete Hermitian manifold with non-positive second Ricci curvature and bounded torsion, and N be a complete Hermitian manifold with its sectional curvature $\leqslant A^2$ and self-torsion $< \tau$. Suppose that $\rho_0 < \dfrac{1}{A} t_g^{-1} \dfrac{A}{\tau}$. If $f \in \mathrm{Hol}(M,N)$ and $f(M) \subset B_{\rho_0}(z)$, then f must be a constant.*

§7. Chen and Yang [15] also gave an estimate of the upper bound for the Levi form of the distance function on the complete Hermitian manifold by using the variational method. With the help of the estimate standard results in Riemannian geometry were extended to Hermitian geometry. The theorem of Bonnet-Myers, the theorem of Y. Tsukamoto [16] etc. are among such extensions.

Two theorems of Chen [15] should be mentioned here.

THEOREM. *Suppose M is a complete Hermitian manifold, whose quasi-holomorphic sectional curvature $\geqslant A^2$. Then the diameter of $M \leqslant \dfrac{\pi}{A}$.*

The quasi-holomorphic sectional curvature decided by vector X is represented by $k'(X)$, then $k'(X) = k(X) - 2|\tau(X)|^2$, where $k(X)$ is ordinary holomorphic sectional curvature decided by X, τ is the torsion form of the Hermitian manifold M. Therefore if M is a Kähler manifold whose holomorphic sectional curvature $\geqslant A^2$, then the diameter of $M \leqslant \dfrac{\pi}{A}$.

THEOREM. *Let M be a complete Hermitian manifold with its quasi-holomorphic sectional curvature satisfying*

$$k'(X)(x) \geqslant \left(\frac{1}{4} + B^2\right) \frac{1}{1 + \rho(x)^2} B > 0$$

where $\rho(x)$ is the distance from the pole $0 \in M$ to x. Then M is compact with its diameter $\leqslant 2(e^{\pi/B} - 1)$.

§8. Using the curvature condition to decide the Steinness of the complex manifold was first studied by R. E. Greene and H. Wu [17]. They proved that a simply connected complete Kähler manifold of non-positive curvature is a Stein manifold. Chen and Yang [18], [19] improved this as follows.

THEOREM. *Suppose N is a Hermitian manifold with a pole p, whose radical curvature $\leqslant \dfrac{1}{4\rho(x)^2}$ and $|T(x)| \leqslant \dfrac{1}{2\rho(x)}$, where $\rho(x) = \mathrm{dist}(p, x)$ and T is the torsion of the Hermitian manifold N. Then N is a Stein manifold.*

The key point of the proof of the theorem is to give a precise estimate of the lower bound of the Levi-form of the square of the distance function $\rho(x)^2$, under the given conditions on curvature and torsion. Then, using the fact that $L\rho^2$ is always positive on the whole manifold, according to a result due to Docquier and Grauert [20], N is a Stein manifold. Note also that for any $a < +\infty$, one can always endow N with a new Hermitian metric so that $B_a(p)$ has holomorphic bisectional curvature bounded from above by a negative constant, so that any holomorphic map from a Kähler manifold of non-negative Ricci curvature into a bounded domain of N must be constant.

§9. In the uniformization theorem of several complex variables, a well-known conjecture is that any compact Kähler manifold X with non-negative holomorphic bisectional curvature and positive Ricci curvature is biholomorphic to a compact Hermitian symmetric space. Mok and Zhong [21] proved this conjecture under the additional assumption that X is Kähler-Einstein. Specifically, they proved the theorem that any compact Kähler-Einstein manifold X with non-negative holomorphic bisectional curvature and positive Ricci curvature is isometric to compact Hermitian symmetric space.

Under the given conditions of the theorem, the main part of the proof of this theorem is to prove that $\nabla R \equiv 0$. The starting point is the method of M. Berger [22]. First of all, if $\alpha \in T^{1,0}_{x_0}(M)$ is a unit vector such that the holomorphic sectional curvature attains its maximum at (x_0, α), by using the maximum principle, one can show that the global maximum of holomorphic sectional curvatures must be attained at every point $x \in X$. The proof of the vanishing of all the terms of $\nabla_\alpha R_{i\bar{j}k\bar{l}}$ is quite tedious and technical.

Then using the invariant property of R under the parallel transport in the direction of the maximal vector, one shows that the curvature tensor is reducible at each point and that the vector space V_x, $x \in X$ consists of the maximal vector and constitutes a differentiable vector bundle invariant under parallel transport. Furthermore, note that the foliation of X by integral submanifolds of the distribution $x \to \Re V_x$ actually corresponds to a global decomposition of X up to a finite covering. This allows one to prove the theorem inductively.

Recently Mok [23] affirmatively resolved the conjecture cited above without any additional assumptions.

Lu [24] studied the relation between holomorphic sectional curvature and sectional curvature and found that for any Kähler manifold M with non-negative (non-positive) sectional curvature at every point the maximum (minimum) of the sectional curvature is attained by the holomorphic sectional curvature.

§10. For a compact Kähler manifold, the problem often arises is that if it could admit some Kähler-Einstein metric. The problem is equivalent to solving the complex Monge-Ampère equation

$$(*) \qquad \det\left(g_{i\bar{j}} + \frac{\partial^2 \phi}{\partial z_i \partial \bar{z}_j}\right) = \det\left(g_{i\bar{j}}\right) e^{F-\phi}, \quad \left(g_{i\bar{j}} + \frac{\partial^2 \phi}{\partial z_i \partial \bar{z}_j}\right) > 0$$

where $(g_{i\bar{j}})$ is the Kählerian metric and $F \in C^\infty(M, \mathbb{R})$ is a given function.

We can solve this equation by the variational method. The difficulty is to obtain *a priori* estimates. Not much important breakthrough had been made until Yau [25] solved the Calabi conjecture, which asserts that every form representing the first Chern class $c_1(M)$ is the Ricci form of some Kähler metric on a compact Kähler manifold (M, g). In fact, Yau proved the following important theorems:

THEOREM F. *Let ϕ be the solution of the equation*

$$(*_t) \qquad \det\left(g_{i\bar{j}} + \frac{\partial^2 \phi}{\partial z_i \partial \bar{z}_j}\right) = \det\left(g_{i\bar{j}}\right) e^{F-t\phi}, \quad \left(g_{i\bar{j}} + \frac{\partial^2 \phi}{\partial z_i \partial \bar{z}_j}\right) > 0.$$

Then

$$0 < m + \Delta\phi \leqslant C_1 \exp\left[C\left(\phi - \left(1 + \frac{t}{m-1}\right)\inf\phi\right)\right]$$

where $m = \dim_{\mathbb{C}} M$, C is a constant such that $C + \inf_{i \neq \ell} R_{i\bar{i}\ell\bar{\ell}} > 1$, and C_1 relies on $\sup_M(-\Delta F)$, $\sup_M|\inf_{i \neq \ell} R_{i\bar{i}\ell\bar{\ell}}|$, $C \cdot m$ and $\sup_M F$.

THEOREM G. *Let ϕ be the solution of $(*_t)$. Then we have an upper bound of $\sum |\phi_{ij\bar{k}}|^2$ by $g_{i\bar{j}}dz^i \otimes d\bar{z}^j$, $\sup|F|$, $\sup|\nabla F|$, $\sup_M \sup_i |F_{i\bar{i}}|$, $\sup_M \sup_{i,j,k}|F_{ij\bar{k}}|$, and $\sup_M|\phi|$.*

In view of this, Yau solved the complex Monge-Ampère equation by continuity method in the case of $c_1(M) < 0$ and $c_1(M) = 0$ in Aubin [26]. But in the case $c_1(M) > 0$, the problem is not yet solved. It is more difficult than the two preceding cases. Therefore the work of Tian Gang [27] is a celebrated one.

Tian's method is a kind of continuity method: let

$$S = \{t \in [0, 1] \,|\, (*_s) \text{ has solutions for } s \in [0, s]\}$$

then by the Yau's work on Calabi conjecture we know $S \neq \emptyset$, and by Aubin [26] we know S is open. In order to prove S is closed which implies $(*)$ is solvable, by using Yau's estimates on the third derivatives of mixed type, one must obtain a C^0 *a priori* estimate of the equation $(*_t)$, Tian introduced some biholomorphic invariants based on the following theorem of L. Hörmander [28].

THEOREM H. *Let $B_R(0)$ be the ball of $\mathbb{C}^n$ with center 0 and radius r and λ be a fixed positive number. Then for all plurisubharmonic functions ψ on $B_R(0)$ with $\psi(0) \geq -1$, $\psi(z) \leq 0$, one has*

$$(**) \qquad \int_{|z|<r} e^{-\lambda \psi(z)} dx \leq C, \quad r < Re^{-\frac{\lambda}{2}}$$

where C is constant relying on m, λ, R and λ is sufficient small.

Let

$$P(M,g) = \left\{ \phi \in C^2(M, \mathbb{R}) \,\bigg|\, \left(g_{i\bar{j}} + \frac{\partial^2 \phi}{\partial z_i \partial \bar{z}_j} \right) \geq 0, \sup_M \phi = 0 \right\}.$$

Using theorem H, Tian defines the following biholomorphic invariants

$$\alpha(M, G) = \sup \left\{ \lambda > 0 \,\big|\, \exists C > 0 \text{ s.t. } (**) \text{ holds for } \phi \in P(M,g) \right\} > 0,$$

The main theorem of Tian is

THEOREM I (TIAN GANG). *Let (M,g) be Kähler manifold. $\frac{1}{\pi}\omega_g$ represents the first Chern class $c_1(M)$ of M. If $\alpha(M) > \frac{m}{m+1}$ where m is the complex dimension of M, then M admits a Kähler-Einstein metric.*

As mentioned above, we must obtain a C^0 *a priori* estimate of ϕ_t. Setting

$$I(\phi) := \frac{(-\sqrt{-1})^m}{V} \int_M \phi(\omega_0^m - \omega^m)$$

$$J(\phi) := \int_0^1 \frac{I(s\phi)}{s} ds$$

where $\omega_0 = g_{\alpha\bar{\beta}} dz^\alpha d\bar{z}^\beta$, $\omega = \omega_0 + \partial\bar{\partial}\phi$, $V = (-\sqrt{-1})^m \int_M \omega_0^m$, $\phi \in P(M,g)$, $\sqrt{-1}\omega > 0$, $\sqrt{-1}\omega_0 > 0$. Then

$$\frac{d}{dt}(I(\phi_t) - J(\phi_t)) = -\frac{(\sqrt{-1})^m}{V} \int_M \phi_t (\Delta_{\phi_t} \dot{\phi}_t)(\omega_0 + \sqrt{-1}\partial\bar{\partial}\phi_t)^m$$

where Δ_{ϕ_t} is the Laplacian of the metric $\left(g_{\alpha\bar{\beta}} + \frac{\partial^2 \phi}{\partial z_\alpha \partial \bar{z}_\beta} \right)$, $\dot{\phi}_t = \frac{d\phi_t}{dt}$. Thus

$$\frac{d(I(\phi_t) - J(\phi_t))}{dt} \geq 0.$$

By passing an elementary and tedious computation, the inequalities

(a) $\qquad \dfrac{(\sqrt{-1})^m}{V} \displaystyle\int_M (-\phi_t)\omega_t^m \leq m \sup_M \phi_t + C, \quad (C \text{ is independent of } \phi_t)$

and
$\forall \varepsilon > 0$, there exists C_ε such that

(b)
$$\sup_M \phi_t \leqslant (m+\varepsilon)\frac{(\sqrt{-1})^m}{V}\int_M (-\phi_t)\omega_t^m + C_\varepsilon$$

are obtained. In order to estimate $\sup_M \phi_t$, we need only estimate

$$\frac{(\sqrt{-1})^m}{V}\int_M (-\phi_t)\omega_t^m.$$

Suppose $\alpha \in \left(\dfrac{m}{m+1}, \alpha(M,g)\right)$. Then by the result of L. Hörmander

$$\int_M e^{-\alpha(\phi_t - \sup_M \phi_t)} dV_M \leqslant C$$

and using Monge-Ampère equation $(*_t)$ and the inequalities (a), (b) we have

$$\frac{(\sqrt{-1})^m}{V}\int_M (-\phi_t)\omega_t^m \leqslant m\frac{1-\alpha}{\alpha}\int_M (-\phi_t)\frac{(\sqrt{-1}\omega_t)^m}{V} + C.$$

Since $m\dfrac{1-\alpha}{\alpha} < 1$, we obtained

$$\frac{(\sqrt{-1})^m}{V}\int_M (-\phi_t)\omega_t^m \leqslant C$$

and at the same time, the *a priori* estimate of $\sup_M \phi_t$ is obtained.

It is not as difficult as in the case of estimate $\sup_M \phi_t$ to obtain an estimates of $\sup_M(-\phi_t)$.

Ding [29] noticed that it could have the following Moser type inequalities

$$\int_M e^{-u} d\mu \leqslant C \exp\left[\eta J(u) - \frac{1}{V}\int_M iu \, d\mu\right]$$

where η, C are positive numbers and $J(u)$ is Aubin-Functional, $u \in P(M,g)$. He defined the following biholomorphic invariants

$$\eta(M) := \inf\left\{\eta(M,g) \mid \omega g \in c_1(M)\right\}$$

and proved the following theorem:

THEOREM. *Suppose M be a compact Kähler manifold with $c_1(M) > 0$. If $\eta(M) < 1$, then M admits Kähler-Einstein metric.*

Ding also pointed that if $\alpha(M,g)$ defined by Tian satisfies $\alpha(M,g) > \dfrac{m}{m+1}$, then $\eta(M) < 1$ is valid. Thus Ding's assumptions is weaker than that of Tian. But the main disadvantage of above results is that few examples have been known about the Kähler manifolds with $\alpha(M,g) > \dfrac{m}{m+1}$ or $\eta(M) < 1$.

Using the method developed by Tian that we have just mentioned above and some estimates of $\alpha(M,g)$, Tian and Yau [30] proved that complex surfaces with differential type $\mathbb{CP}^2 \# n\overline{\mathbb{CP}^2}$ $(3 \leqslant n \leqslant 8)$ and positive first Chern class admit Kähler-Einstein metric.

§11. The famous Frankel Conjecture asserts that every compact Kähler manifold with nonnegative sectional curvature is biholomorphic to $\mathbb{CP}^n$. This conjecture has been proved by Siu-Yau [31] and Mori [32] in a general case. In view of this, one would consider the corresponding problems in a negative-curvature case. Siu and Yang [33] proved the following theorem:

THEOREM. *Suppose M be a compact Kähler-Einstein surface with nonpositive holomorphic bisectional curvature. Suppose $\chi < \dfrac{2}{3(1+\sqrt{\frac{6}{11}})}$,*

$$K_{ar}(P) - K_{\min}(P) \leqslant \chi(K_{\max}(P) - K_{\min}(P))$$

where $P \in M$. Then M is biholomorphic isomorphic equivalent to the compact quotient of a complex 2-ball with invariant metric.

The method of Siu and Yang is to consider the function

$$\phi = 4(3r_2 - r_1^2) - 3(3K_{\min}(P) - 2\rho)^2,$$

where r_1, r_2 are the Chern-Weil function of the curvature tensor of M and ρ is the scalar curvature. Then ϕ is a global function defined on M. Siu and Yang obtained $\Delta\phi^\lambda(P) < 0$ on the non-ball like point P of M for a sufficient small positive number λ. On the other hand, the real dimension of ball like points will be less than 2. Thus ϕ^λ can smoothly be extended to M and must be a constant 0 everywhere.

Y. Hong, Z. D. Guan, and H. C. Yang [33] noticed that λ can be chosen to be $\dfrac{1}{6}$, and

$$\Delta\phi^{\frac{1}{6}} = \frac{1}{6}\phi^{-\frac{11}{6}}\left(\phi\,\Delta\phi - \frac{5}{6}\sum_\delta |\nabla_\delta\phi|^2\right)$$

where

$$\sum_\delta |\nabla_\delta\phi|^2 = 36(|B\nabla_1 R_{1\bar{2}1\bar{2}} + (A^2 - B^2)\nabla_2\rho^2|^2$$

$$+ B\nabla_{\bar{2}} R_{1\bar{2}1\bar{2}} - (A^2 - B^2)\nabla_{\bar{1}}\rho^2|^2),$$

$A = 2R_{1\bar{1}2\bar{2}} - R_{1\bar{1}2\bar{1}}$, $B = R_{1\bar{2}1\bar{2}}$ and $K_{\min}(P) = R_{1\bar{1}1\bar{1}} = R_{2\bar{2}2\bar{2}}$. Since

$$5|B\nabla_1 R_{1\bar{2}1\bar{2}} + (A^2 - B^2)\nabla_2\rho^2|^2 + 5(6B^2 - A^2)(A^2 - B^2)|\nabla_2\rho^2|^2$$
$$= (A^2 - B^2)|\nabla_1 R_{1\bar{2}1\bar{2}} + 5B\nabla_2\rho^2|^2 + (6B^2 - A^2)|\nabla_1 R_{1\bar{2}1\bar{2}}|^2$$
$$\geqslant 6(B^2 - A^2)|\nabla_1 R_{1\bar{2}1\bar{2}}|^2,$$

therefore

$$\Delta\phi^{\frac{1}{6}} \leqslant -5\phi$$

and ϕ must be 0 everywhere. Thus one can conclude from this that the constant χ can be chosen to $\chi < \dfrac{2}{3(1 + \sqrt{\frac{1}{6}})}$ which is better than the constant obtained by Siu and Yang.

§12. W.-X. Shi [34, 35] uses the evolution equation to investigate the deformation of the Ricci curvature of a given Riemannian manifold which was first introduced by R. S. Hamilton [36]. The equation is:

$$(***) \qquad \frac{\partial g_{ij}}{\partial t} = -2R_{ij}$$

where (M, g_{ij}) is a given Riemannian manifold and R_{ij} its Ricci tensor. The first problem naturally arises is the short time existence of the solution of the equation. Hamilton proved this for M being compact. However, the noncompact case is much more difficult than the compact case. The main difficulty in solving the above equation comes from the fact that the equation is not strictly parabolic but weak parabolic. Shi modified the equation $(***)$ to

$$(***_s) \qquad \frac{\partial}{\partial t} g_{ij} = -2R_{ij} + \nabla_i V_j + \nabla_j V_i,$$

which is a strictly parabolic equation. He considered the Dirichlet problem of this equation on the exhausted compact region of the complete noncompact manifold. By an estimate of the derivative of the curvature tensor, Shi proved the following theorem:

THEOREM. *Let $(M, g_{ij}(x))$ be an n-dimensional complete noncompact Riemannian manifold with its Riemannian curvature tensor $\{R_{ijk\ell}\}$ satisfying*

$$|R_{ijk\ell}|^2 \leqslant k_0, \quad on \ M,$$

where $0 < k_0 < +\infty$ is a constant. Then there exists a constant $T(n,k) > 0$ depending only on n and k such that the evolution equation

$$\frac{\partial}{\partial t} g_{ij}(x,t) = -2R_{ij}(x,t), \quad on \ M$$
$$g_{ij}(x,0) = g_{ij}(x), \quad \forall x \in M$$

has a smooth solution $g_{ij}(x,t) > 0$ for a short time $0 \leqslant t \leqslant T(n, k_0)$ ans satisfies the following estimates: for any integer $m \geqslant 0$ there exist constants C_m depending only on n, m and k_0 such that

$$\sup_{x \in M} |\nabla^m R_{ijk\ell}(x,t)|^2 \leqslant C_m / t^m, \quad 0 < t \leqslant T(n, k_0)$$

By developing a much more complicated maximal principle of the heat equation on noncompact manifolds and long time existence theorem, Shi also proved the following

THEOREM. *Let M be an n-dimensional complete noncompact Riemannian manifold, $n \geqslant 3$. For any $C_1, C_2, \delta > 0$, there exists a constant $\varepsilon = \varepsilon(n, C_1, C_2, \delta)$ such that the curvature of M satisfies*

(a) $\operatorname{Vol}(B(x, r)) \geqslant C_1 r^n \quad \forall x \in M, r \geqslant 0,$

(b) $|\overset{o}{Rm}|^2 < \varepsilon R^2, \quad 0 \leqslant R \leqslant C_2 / r(x_0, x)^{\varepsilon + \delta}, \quad \forall x \in M,$

where $|\overset{o}{Rm}|^2 = |W_{ijk\ell}|^2 + |V_{ijk\ell}|^2$ and $W_{ijk\ell}$ and $V_{ijk\ell}$ are Weyl conformal curvature tensor and $V_{ijk\ell}$ the traceless Ricci part of the curvature tensor. Then the evolution equation

$$\begin{cases} \dfrac{\partial}{\partial t} g_{ij}(x, t) = -2R_{ij}(x, t) \\ g_{ij}(x, 0) = g_{ij}(x) \end{cases}$$

has a solution for all time $0 \leqslant t < \infty$ and the metric $g_{ij}(t)$ converges to a smooth metric $g_{ij}(\infty)$ as time $t \to +\infty$ such that $R_{ijk\ell}(\infty) \equiv 0$ on M.

Based on above results, Shi [37] provided yet another uniformization theorem:

THEOREM. *Suppose M is a complete noncompact Kähler manifold of complex dimension n with bounded and positive holomorphic bisectional curvature. Suppose M satisfies the following additional assumptions:*

(a) $\operatorname{Vol}(B(x_0, r)) \geqslant C_0 r^{2n}, \quad 0 \leqslant r \leqslant +\infty,$

(b) $\int_{B(x_0, r)} R(x)dx \leqslant C_1 r^{2n-2}, \quad x_0 \in M, \quad 0 \leqslant r \leqslant +\infty,$

where $0 < C_0, C_1 < +\infty$ are some constants. Then M is biholomorphic equivalent to $\mathbb{C}^n$.

REFERENCES

1. Carathéodory, C., Math. Ann. **72** (1912), 107–144.
2. Pick, G., Math. Ann. **77** (1916), 7–13.
3. Lu, Q. K., Advance in Math. (China) **7** (1957), 370–420. (Chinese)
4. Lu, Q. K., *Classical Domains and Classical Manifolds*, Shanghai Science and Technology Press, 1965. (Chinese)
5. Ahlfors, L., Trans. A.M.S. **43** (1938), 359–364.
6. Yau, S. T., Amer. J. of Math. **100** (1978), 197–203.
7. Chen, Z. H., Cheng, S. Y. and Lu, Q. K., Scientia Sinica **22** (1979), 1238–1247.
8. Lu, Q. K., Scientia Sinica **7** (1958), 453–504.
9. Yang, H. C., Chen, Z. H., Acta Math. Sinica (China) **24(6)** (1981), 945–952.
10. Chen, Z. H. and Yang, H. C., Bulletin in Sciences **28(6)** (1983), 879–882.
11. Goldberg, S. I. and Ishihara, T., Amer. J. of Math. **98** (1976), 225–240.
12. Goldberg, S. I. and Har'El, Z., Bull. Greek Math. Society **18** (1977), 141–148.
13. Yang, H. C. and Chen, Z. H., Proceedings of the 1981 Hongzhou Conference, Birkhäuser Boston Inc., 1984, pp. 99–116.
14. Chen, Z. H. and Yang, H. C., Acta Math. Sinica **28(2)** (1985), 218–232.
15. Chen, Z. H., Acta Math. Sinica **27(5)** (1984), 631–643. (Chinese)
16. Tsukamoto, Y., Proc. Japan Acad. **33** (1957), 333–335.
17. Greene, R. E. and Wu, H., Lecture Notes in Mathematics, No. 699, Springer-Verlag, 1979.
18. Chen, Z. H. and Yang, H. C., Scientia Sinica (Series A) **27(5)** (1984), 457–460.
19. Chen, Z. H., Lecture Notes in Mathematics, No. 1255, Springer-Verlag, 1987, pp. 13–25.
20. Docqier, F. and Grauert, H., Math. Ann. **140** (1960), 94–123.
21. Mok, Mgaiming and Zhong, J. Q., J. Diff. Geom. **23** (1986), 15–67.

22. Berger, M., CR. III, Reunion Math. Expression, Latine (Namur, 1965), 35–55.
23. Mok, Ngaiming, Lecture Notes in Mathematics, No. 1194, Springer-Verlag, 1986.
24. Lu, Q. K., Scientia Sinica **26** (1987), 787–793.
25. Yau, S. T., Comm. Pure. Appl. Math. **31** (1978), 339–411.
26. Aubin, T., *Nonlinear Analysis on Manifolds. Monge-Ampère Equations*, Grundlehren der mathematischen wissenschaften, 252, Springer-Verlag, 1982.
27. Tian, G., Invent. Math. **89** (1987), 225–246.
28. Hörmander, L., *An Introduction to Several Complex Variables*, Van Nostrand, Princeton, NJ, 1973.
29. Ding, W. Y., Math. Ann. **282** (1988), 463–471.
30. Tian, G. and Yau, S. T., Comm. Math. Phy. **112** (1987), 175–203.
31. Siu, Y. T. and Yau, S. T., Invent. Math. **59** (1980), 189–204.
32. Siu, Y. T. and Yang, P., Invent. Math. **64** (1981), 471–487.
33. Hong, Y., Guan Z. D. and Yang, H. C., Acta Math. Sinica **31** (1988), 595–602. (Chinese)
34. Shi, W. X., J. Diff. Geom. **30** (1989), 223–301.
35. Shi, W. X., J. Diff. Geom. **30** (1989), 303–394.
36. Hamilton, R. S., J. Diff. Geom. **17** (1982), 225–306.
37. Shi, W. X., Bull. Amer. Math. Soc. **30** (1990), 433–436.

DEPARTMENT OF APPLIED MATHEMATICS, SHANGHAI JIAOTONG UNIVERSITY, SHANGHAI 200030, PEOPLE'S REPUBLIC OF CHINA

Contemporary Mathematics
Volume **142**, 1993

Biholomorphic Mappings in
Several Complex Variables

SHENG GONG

Contents

1991 *Mathematics Subject Classification.* 32H02, 32A30, 30C35, 32-02.
Research supported by the National Science Foundation of China.
This paper is in final form and no version of it will be submitted for publication elsewhere.

Introduction

It is well known that there are many fundamental and interesting results in geometric function theory of one complex variable that are generally not true in versions of several complex variables. H. Cartan [CarH] is the first mathematician to systematically extend geometrical function theory from one variable to several variables. In 1933, in P. Montel's book on univalent function theory, Henri Cartan [CarH] wrote an appendix entitled *Sur la possibilite d'entendre aux fonctions de plusieurs variables complexes la theorie des fonctions univalents* in which he called for a number of generalizations of properties of univalent functions in one variable to biholomorphic mappings in several variables. He pointed out that there does not exist the corresponding Bieberbach conjecture in the case of several complex variables even in the simplest situation and that a boundedness of the modulus of the second coefficients of the Taylor expansions of the normalized univalent functions on the unit disc is not true in the several complex variables case. He also pointed out that the corresponding growth and covering theorems fail in the case of several complex variables. He indicated particular interest in the properties of the determinant of the complex Jacobian of biholomorphic mappings in several complex variables (the square of the magnitude of the determinant of the complex Jacobian is the infinitesimal magnification factor of volume in $\mathbb{C}^n$). He stated a *theoreme presume* that the magnitude of the determinant of the Jacobian of a normalized biholomorphic mapping would have a finite upper and a positive lower bound depending only on $|z| = r < 1$. He also illustrated the significances and merits in determining these bounds. That his conjecture does not hold has been known for some time. For demonstration, we exhibit the following counter-example. For any positive integer k, let $f(z) = (f_1(z), f_2(z))$ with

$$\begin{cases} f_1(z) = z_1, \\ f_2(z) = z_2(1 - z_1)^{-k} = z_2 + kz_2z_1 + \cdots, \end{cases}$$

then f is a normalized biholomorphic function on the unit ball B^2 in $\mathbb{C}^2$, that is $f(0) = 0$, the Jacobian J_f of f at $z = 0$ is identity matrix and J_f is given by

$$J_f = \begin{bmatrix} 1 & 0 \\ \dfrac{kz_2}{(1 - z_1)^{k+1}} & \dfrac{1}{(1 - z_1)^k} \end{bmatrix}.$$

Thus $|\det J_f| = |1 - z_1|^{-k}$, which yields

$$\max_{|z| \leqslant r} |\det J_f| = (1 - r)^{-k} \to \infty \qquad \text{as } k \to \infty,$$

and

$$\min_{|z| \leqslant r} |\det J_f| = (1 + r)^{-k} \to 0 \qquad \text{as } k \to \infty.$$

In the same paper H. Cartan suggested the study of starlike mappings and convex mappings and some other important classes of mappings in several complex variables.

Since then many mathematicians have worked on this field and derived many significant results.

Since 1988, there have been many attempts in systematically extending the geometric function theory of one complex variable to several complex variables. The present article surveys some of the results obtained by the author and relevant researchers in the past few years. Here we shall announce the statements only, because most the proofs are quite long.

In this survey, we shall focus on results related to bounded symmetric domains as well as Reinhardt domains. In section I, it will be the conditions of starlikeness, convexity and univalency for holomorphic mappings. In section II, it will be the consideration of the distortion theorems of family of $\mathcal{A}$-invariant mappings, and the presentation of some important results in this respect. In section III, we present two different types of growth theorems and covering theorems for starlike mappings and point out the equivalency of these two different types of growth theorems. In section IV, the distortion theorem, growth theorem and covering theorem of convex mappings on classical domains are given. Also, the precise form of the distortion theorem in matrix form of biholomorphic convex mappings of ball is stated. How to establish the Schwarzian derivative in several complex variables has been an interesting problem. In section V, we try to define the Schwarzian derivatives of holomorphic mappings in high dimensions. In the last section of this article, it is about the Bloch constant of bounded symmetric domains, which seems to be another interesting problem in several complex variables. Finally an estimation of the bounds of the Bloch constant on bounded symmetric domains will be given.

Of course we have touched only a few problems in this field, and obtained some partial results. A lot of interesting problems remains to be resolved.

I would like to take this opportunity to express my sincerely thanks to the Department of Mathematics, University of California, San Diego, particularly Professor Carl H. FitzGerald; and to Department of Mathematics and Statistics, Southern Illinois University at Edwardsville, particularly Professor Chung-Wu Ho, for their hospitalities in providing me with stimulating environments, where some of the researches reported here were carried out.

I. The conditions of starlikeness, convexity and univalency for a holomorphic mapping

1. The necessary and sufficient conditions of starlikeness for a holomorphic mapping on Reinhardt domain D_p.

Throughout this paper a starlike mapping means that the image of the mapping is starlike with respect to the origin.

In 1970, T. J. Suffridge [**S1**] established the necessary and sufficient conditions of starlikeness for a holomorphic mapping on Reinhardt domain

$$B_p = \left\{ z = (z_1, \ldots, z_n) \in \mathbb{C}^n \;\middle|\; \sum_{j=1}^n |z_j|^p < 1 \right\},$$

where p is a real number $\geqslant 1$. It can be stated as follows. Suppose $f : B_p \to \mathbb{C}^n$ is locally biholomorphic with $f(0) = 0$. Then f is starlike if and only if $f(z) = J_f(z)w$, where $w \in \mathcal{P}_p$ and $\mathcal{P}_p$ is a family of holomorphic mapping; $w \in \mathcal{P}_p$ if $w : B_p \to \mathbb{C}^n$, $w(0) = 0$ and $\operatorname{Re} \sum_{j=1}^n w_j |z_j|^p / z_j \geqslant 0$ holds for all $z \in B_p$ (where $f(z)$ and w are column vectors).

K. Kikuchi [**Kik**] discussed the necessary and sufficient condition of starlikeness for holomorphic mappings on bounded domains. He used the Bergman kernel function to express the condition. In the case of the unit ball, the conditions of Kikuchi and that of Suffridge coincide with each other.

Gong, Wang and Yu [**GWY3**] have extended the Suffridge's result to the Reinhardt domain D_p, where

$$D_p = \left\{ z = (z_1, \ldots, z_n) \in \mathbb{C}^n \; \middle| \; \sum_{i=1}^n |z_i|^{p_i} < 1, \; 2p_n > p_1 \geqslant p_2 \geqslant \cdots \geqslant p_n > 1 \right\}.$$

Suppose $f : D_p \to \mathbb{C}^n$ is locally biholomorphic with $f(0) = 0$. Then f is star-like if and only if $\langle f^{-1*}du, d\rho \rangle|_{w=f(z)} \geqslant 0$ holds for any $z \in D_p$, where $\langle , \rangle$ means the inner product defined in $\mathbb{C}^n$, $u(z) = \sum_{i=1}^n |z_i|^{p_i}$, $\rho(w) = \left(\sum_{i=1}^n |w_i|^2 \right)^{1/2}$, $w = (w_1, \ldots, w_n) = f(z)$, f^{-1*} means pull back of f^{-1} on U_z where U_z is a neighborhood of z such that $f(z)$ is biholomorphic there.

When $p_1 = p_2 = \cdots = p_n > 1$, the foregoing is just the condition set by Suffridge.

In proving this theorem, a Schwarz type lemma on D_p is needed. It can be stated as follows.

If ϕ is a holomorphic mapping with $\phi(0) = 0$, $J_\phi(0) = \mu I$, $0 < \mu < 1$, which maps D_p into D_p, then $u(\phi(z)) \leqslant u(z)$ holds for any $z \in D_p$ where $u(z) = \sum_{i=1}^n |z_i|^{p_i}$.

C. H. FitzGerald pointed out that the above Schwarz type lemma does not hold if the condition $2p_n > p_1$ fails to hold. The following counter-example is given: Let $D = \{ (z_1, z_2) \in \mathbb{C}^2 \mid |z_1|^5 + |z_2|^2 < 1 \}$, $\phi(z_1, z_2) = (\varepsilon z_1, \varepsilon z_2 + (1 - 2\varepsilon)z_1^2)$ and ε is sufficiently small.

2. The necessary and sufficient conditions of starlikeness for a holomorphic mappings on Caratheodory complete domains.

Gong, Wang and Yu [**GWY4**] have extended the result of the previous section further to more general domains.

If Ω is a domain in $\mathbb{C}^n$, $r : \Omega \to \mathbb{R}^+ + \{0\}$ belongs to C^1 class, and $f : \Omega \to \mathbb{C}^n$ is a locally biholomorphic mapping on Ω and satisfies the following conditions:

   i) $r(p) = 0$ for a fixed $p \in \Omega$; and $r(z) > 0$ when $z \in \Omega - \{p\}$,

   ii) $f(p) = 0$,

   iii) for any $z \in \Omega$, there exists a neighborhood U_z of z, such that f is biholomorphic on U_z (such a neighborhood exists since f is locally biholomorphic on Ω) and $\langle f^{-1*}dr, d\rho \rangle|_{w=f(z)} \geqslant 0$ holds for any $z \in \Omega$, where $\langle , \rangle$ means the inner product of $\mathbb{C}^n$, $\rho = \left(\sum_{i=1}^n |w_i|^2 \right)^{1/2}$, $w = (w_1, \ldots, w_n) = f(z)$, and f^{-1*} means the pull back of f^{-1} on U_z,

then f is biholomorphic on Ω, and $f(\Omega)$ is starlike with respect to $f(p) = 0$.

In fact, we can derive a more general result as follows.

Suppose Ω is a domain in $\mathbb{C}^n$, $r : \Omega \to \mathbb{R}^+ + \{0\}$ is a continuous exhaustion function which satisfies the conditions i), ii) and

   iii') for any $z \in \Omega$, there exists a neighborhood U_z, such that $f(z)$ is biholomorphic on U_z and for any $1 > t > 0$, $(1 - t)f(z) \in f(U_z)$, $f(z_t) = (1 - t)f(z)$, $z_t \in U_z$, the inequality $r(z_t) \leqslant r(z)$ holds.

Then the above conclusions still hold.

It is of interest to apply these results to the Caratheodory complete domains.

Suppose $\Omega \in \mathbb{C}^n$ is a domain, $p, q \in \Omega$, and the Caratheodory pseudo-distance of p, q is defined as

$$C_\Omega(p, q) = \sup_{\substack{f \in \mathcal{O}(\Omega, D) \\ f(p)=0}} \frac{1}{2} \log \frac{1 + |f(q)|}{1 - |f(q)|},$$

where $D = \{\xi \in C \mid |\xi| < 1\}$ is the unit disc, $\mathcal{O}(\Omega, D)$ is the family of all holomorphic functions which maps Ω into D. It can be shown that $C_\Omega(p, q)$ defines a distance function if Ω is bounded.

The domain Ω is Caratheodory complete if for any $p \in \Omega$ and any $a \in \mathbb{R}^+$, the set $\{z \in \Omega \mid C_\Omega(p, z) < a\}$ is relative compact on Ω.

Poletskii and Shabat [PS] pointed out: Strict pseudoconvex domains are Caratheodory complete domains. Kobayashi [Kob] proved that the general analytic polyhedrons are Caratheodory complete domains. Moreover, Kim [Kim] showed that if $D \subset \mathbb{C}^n$ is a domain, and if there exists a set $K \subset\subset D$, such that for any $z \in D$, there corresponds a biholomorphic automorphism $f \in \operatorname{Aut} D$ with $f(z) \in K$, then D is a Caratheodory complete domain.

In [GWY4], Gong, Wang and Yu as a consequence of the previous results, gave the necessary and sufficient conditions of starlikeness for locally biholomorphic mappings on Caratheodory complete domains.

Suppose $\Omega \in \mathbb{C}^n$ is a Caratheodory complete domain, $0 \in \Omega$, $f : \Omega \to \mathbb{C}^n$ is a locally biholomorphic mapping with $f(0) = 0$, and for any $z \in \Omega$, there exists a neighborhood U_z, such that $f(z)$ is biholomorphic on U_z. Then $f(z)$ is a biholomorphic starlike mapping on Ω if and only if for any $1 > t > 0$, $(1-t)f(z) \in f(U_z)$, $f(z_t) = (1-t)f(z)$, $z_t \in U_z$, and the inequality $C_\Omega(0, z_t) \leqslant C_\Omega(0, z)$ holds.

The four classical domains are defined as (cf. L. K. Hua [Hua]):

$$
\begin{aligned}
&\mathcal{R}_I : I - Z\bar{Z}' > 0, && Z = Z^{(m,n)} = (z_{jk})_{1 \leqslant j \leqslant m, 1 \leqslant k \leqslant n}; \\
&\mathcal{R}_{II} : I - Z\bar{Z} > 0, && Z = Z^{(n)} = (z_{jk})_{1 \leqslant j, k \leqslant n}; \\
&\mathcal{R}_{III} : I + Z\bar{Z} > 0, && Z = -Z' = Z^{(n)} = (z_{jk})_{1 \leqslant j, k \leqslant n}; \\
&\mathcal{R}_{IV} : 1 - 2\bar{z}z' + |zz'|^2 > 0, && |zz'| < 1, z = (z_1, \ldots, z_n) \in \mathbb{C}^n.
\end{aligned}
$$

The norms of the four classical domains are defined as follows (cf. L. A. Harris [Har]): For $\mathcal{R}_I$, $\mathcal{R}_{II}$, $\mathcal{R}_{III}$, the norm $\| \cdot \|_K$ ($K = I, II, III$) is defined to be the positive square root of the largest eigenvalue of $Z\bar{Z}'$; for $\mathcal{R}_{IV}$, the norm $\| \cdot \|_{IV}$ is defined as $(|z|^2 + (|z|^4 - |zz'|^2)^{1/2}$.

For the four classical domains, the Caratheodory distance is defined as:

$$C_{\mathcal{R}_K}(0, z) = \frac{1}{2} \log \frac{1 + \|z\|_K}{1 - \|z\|_K}$$

where $\| \cdot \|_K$ is the norm of the element of $\mathcal{R}_K$, $K = I, II, III, IV$. The Caratheodory distance for the Reinhardt domain

$$B_p = \left\{ z = (z_1, \ldots, z_n) \in \mathbb{C}^n \;\middle|\; \|z\|_p = \left(\sum_{i=1}^n |z_i|^p \right)^{1/p} < 1 \right\}$$

is $C_{B_p}(0, z) = \dfrac{1}{2} \log \dfrac{1 + \|z\|_p}{1 - \|z\|_p}$.

As a corollary, we have the following result. Suppose Ω is any one of the previous domains and $\| \cdot \|$ is the corresponding norm of the domain. If $f : \Omega \to \mathbb{C}^n$ is a locally biholomorphic mapping with $f(0) = 0$, and for any point $z \in \Omega$, there exists a neighborhood U_z, such that $f(z)$ is biholomorphic on U_z, then $f(z)$ is biholomorphic starlike mapping on Ω if and only if for any $1 > t > 0$, $(1-t)f(z) \in f(U_z)$, $f(z_t) = (1-t)f(z)$, $z_t \in U_z$, and the inequality $\|z_t\| \leqslant \|z\|$ holds.

Because all the norms of these five domains belong to C^1 class, so we can restate the corollary as follows. Suppose Ω is any one of the five domains and $\| \cdot \|$ is the corresponding norm of the domain. If $f : \Omega \to \mathbb{C}^n$ is a locally biholomorphic mapping with $f(0) = 0$, and for any point $z \in \Omega$, there exists a neighborhood U_z, such that $f(z)$ is biholomorphic on U_z, then $f(z)$ is biholomorphic starlike mapping if and only if $\langle f^{-1*} d\|z\|, d\rho \rangle|_{w=f(z)} \geqslant 0$, where $\rho = \left(\sum_{i=1}^n |w_i|^2 \right)^{1/2}$, $w = (w_1, \ldots, w_n) = f(z)$, and f^{-1*} means the pull back of f^{-1} on U_z.

3. A sufficient condition of univalency for a locally biholomorphic mapping.

Let $f(z)$ be an analytic function defined in the open unit disc $|z| < 1$ of the complex plane $\mathbb{C}$ which satisfies the condition

$$(1 - |z|^2) \left| \frac{z f''(z)}{f'(z)} \right| \leqslant c.$$

In 1972, J. Becker showed that $c = 1$ implies $f(z)$ is univalent at the unit disc and that $c < 1$ implies f can be extended to quasi-conformal homeomorphism of the whole plane.

In 1974, J. A. Pfaltzgraff [Pfa1] extended the Becker's sufficient condition for univalent of an analytic function in the unit disc of the complex plane to the unit ball in $\mathbb{C}^n$, by first establishing the Loewner chain as defined below.

Let $\{f(z, t)\}$, $t \geqslant 0$ be a family of holomorphic mappings on B^n with $f(0, t) = 0$. We define that $\{f(z, t)\}$ is a subordinate chain if $f(z, s)$ is subordinate to $f(z, t)$ when $0 \leqslant s \leqslant t \leqslant \infty$. In particular, if $f(z, t)$ are biholomorphic mappings, then we call it Loewner chain. Pfaltzgraff established the relation between the Loewner chain and the Loewner differential equation.

By using the Loewner chain J. A. Pfaltzgraff [Pfa1] proved that: If $f(z)$ is a normalized locally biholomorphic mappings on the unit ball B^n in $\mathbb{C}^n$ which maps the unit ball to $\mathbb{C}^n$ and satisfies the condition

$$(1 - |z|^2) \left\| J_f^{-1}(z) \frac{d^2 f}{dz^2}(z)(z, \cdot) \right\| \leqslant c, \qquad 0 \leqslant c \leqslant 1$$

where $\frac{d^2 f}{dz^2}(z)(z, \cdot)$ is a $n \times n$ matrix for which the (i, j) entry is $\sum_{k=1}^n \frac{\partial^2 f_i(z)}{\partial z_j \partial z_k} z_k$, $\| \cdot \|$ is the norm of matrix, then $f(z)$ is a biholomorphic mapping on the unit ball B^n.

Also by using the Loewner chain, Hong-bin Chen and Fu-yao Ren [CR] got some sufficient conditions of univalence for locally biholomorphic mappings. Here we state two of them that generalize the result of Pfaltzgraff.

If $f(z)$ is a normalized locally biholomorphic mapping which maps B^n into $\mathbb{C}^n$, and it satisfies the following condition

$$\left\| (1 - |z|^2) J_f^{-1}(z) \frac{d^2 f}{dz^2}(z)(z, \cdot) + c\|z\|^2 I \right\| < 1,$$

then $f(z)$ is a normalized biholomorphic mapping on B^n, where $|c| \leqslant 1$, $c \neq 1$, I is the identity matrix.

If $f(z)$ is a normalized locally biholomorphic mapping which maps B^n into $\mathbb{C}^n$, and there exists a complex number λ, $\text{Re}\,\lambda > 1$, such that

$$\left\| (1 - |z|^\lambda) J_f^{-1}(z) \frac{d^2 f}{dz^2}(z)(z, \cdot) + (1 - \frac{\lambda}{2}) I \right\| < \frac{|\lambda|}{2},$$

then $f(z)$ is a normalized biholomorphic mapping on B^n.

The results of Pfaltzgraff and Hong-bin Chen and Fu-yao Ren can be extended to the complex Banach space of finite dimensions.

Moreover, Pfaltzgraff got the growth and covering theorems as follows:

$$\frac{|z|}{(1 + c|z|)^2} \leqslant |f(z)| \leqslant \frac{|z|}{(1 - c|z|)^2}$$

and $f(B^n)$ contains $\frac{1}{(1 + c)^2} B^n$ if $f(z)$ satisfies the conditions mentioned above.

In 1990, Pfaltzgraff [Pfa2] pointed out: If one defines some norm $\| \cdot \|$ on $\mathbb{C}^n$ such that $f(z, t)$ is the Loewner chain on $\|z\| < 1$, and $f(z, 0) = f(z)$, then

$$\frac{\|z\|}{(1 + \|z\|)^2} \leqslant \|f(z)\| \leqslant \frac{\|z\|}{(1 - \|z\|)^2}, \qquad z \in B^n.$$

Note that the normalized biholomorphic starlike mappings and the normalized close-to-starlike mappings (cf. Suffridge [S3]) all satisfy these conditions. It follows that we have obtained the corresponding growth theorems and moreover the Koebe covering theorem: $f(B^n)$ contains $\frac{1}{4} B^n$.

4. The necessary and sufficient condition of convexity for holomorphic mappings.

What is the necessary and sufficient conditions of convexity for a biholomorphic mapping?

In 1970, T. J. Suffridge [S2] proved the following result.

Suppose $f(z) = (f_1(z), \ldots, f_n(z))$, $z = (z_1, \ldots, z_n) \in \mathbb{C}^n$ is a biholomorphic convex mapping on polydisc P^n. Then there exists a constant matrix $T = T^{(n)}$, such that $f(z) = (\phi_1(z_1), \ldots, \phi_n(z_n))T$ holds, where $\phi_j(z_j)$ is an analytic convex function of one complex variable z_j on the unit disc $|z_j| < 1$, $j = 1, 2, \ldots, n$. That means, the biholomorphic convex mappings of several complex variables on the polydisc are the direct product (with possible difference by constant matrix) of the analytic convex functions of one complex variable.

K. Kikuchi [Kik] obtained the necessary and sufficient conditions of convexity of holomorphic mappings on bounded domains. However, we can give some counter-examples to show that these conditions are necessary but may not be sufficient. By Gong, Wang and Yu [GWY5], the following result has been proved. If $w = f(z) = (w_1, \ldots, w_n)$ is a locally biholomorphic mapping which maps B^n to $\mathbb{C}^n$ with $f(0) = 0$, then f is biholomorphic convex if and only if: for any $z \in B^n$ and any unit vector $b = (b_1, \ldots, b_n)$ such that $\mathrm{Re}(z\bar{b}') = 0$, then

$$1 + \mathrm{Re}\left(\sum_{k,\ell,m} \sum_{i,j} b_\ell b_m \frac{\partial w_i}{\partial z_\ell} \frac{\partial \bar{w}_j}{\partial z_m} z_k \frac{\partial^2 z_k}{\partial w_i \partial w_j} \right) \geqslant 0.$$

Ngaiming Mok and I-Hsun Tsai [MT] proved the following result about the bounded biholomorphic convex mappings on irreducible Hermitian symmetric manifold of the non-compact type and of rank $\geqslant 2$.

Let X_0 be an irreducible Hermitian symmetric manifold of non-compact type and of rank $\geqslant 2$ and $\tau : X_0 \to \Omega \subset\subset \mathbb{C}^n$ be the Harish-Chandra embedding. Let D be a bounded convex domain in $\mathbb{C}^n$ and $f : X_0 \to D$ be a biholomorphism. Then f is the Harish-Chandra embedding up to automorphisms of X_0 and affine linear transformations of $\mathbb{C}^n$. More precisely, f is of the form $T \circ \tau \circ \phi$, where T is an affine linear transformation of $\mathbb{C}^n$ and ϕ is an automorphism of X_0.

II. Distortion theorem for biholomorphic mappings in transitive domains

1. Distortion theorem for family of $\mathcal{A}$-invariant mappings.

Various distortion theorems for families of univalent functions have been studied since as early as 1907 when Koebe discovered his *Verzerrungsatz* the distortion theorem for the class of univalent functions defined on the unit disc in the complex plane $\mathbb{C}$. Koebe's theorem gives explicit upper and lower bounds on $|f'(z)|$ in terms of $|z|$ for any function f that is one-to-one and analytic on the unit disc and normalized by $f(0) = 0$ and $f'(0) = 1$. The term distortion arises from the geometric interpretation of $|f'(z)|$ as the infinitesimal magnification factor of arc length and the interpretation of the square of $|f'(z)|$ as the infinitesimal magnification factor of area.

We note that the counter-example in the Introduction also applies for z in the polydisc $\{(z_1, z_2) \mid |z_1| < 1, |z_2| < 1\}$ as well as on the unit ball. We also note from this counter-example that there is no bound on the magnitude of second coefficients of normalized biholomorphic mappings on the polydisc or on the ball B^2. The lack of such a bound on these coefficients shows that the usual proofs in one variable of Koebe's distortion theorem cannot be extended to $\mathbb{C}^n$ for $n \geqslant 2$. Also Cartan conjectured that the distortion theorem does not hold in the class of normalized biholomorphic mappings in $\mathbb{C}^n$ $(n \geqslant 2)$ which suggests that the conjecture may be held only for a proper subclass of functions.

As mentioned, the invariance of the full family of univalent functions under Möbius transformations has been a powerful tool in obtaining theorems and other results concerning the subfamily of normalized univalent functions. Similarly, in several complex variables studies, invariance has also played an important role. W. Rudin [R] used the term $\mathcal{M}$-*invariant space* to mean a function space of holomorphic mappings on B^n in $\mathbb{C}^n$ that is invariant under Möbius transformations. Similarly, he used $\mathcal{U}$-*invariant space* to mean a space that is invariant under unitary transformations. We shall say that a space is $\mathcal{A}$-*invariant* if it is invariant under the full group of holomorphic automorphisms of B^n. For example, the family of biholomorphic mappings, the family of biholomorphic convex mappings, the family of local biholomorphic mappings and the family of local biholomorphic convex mappings are families of $\mathcal{A}$-invariant holomorphic mappings. For one complex variable, Ch. Pommerenke [Pom1, Pom2] discussed such families of analytic functions. He used the name *Linear invariant family*, and obtained several interesting results about such families in the cited references and the articles that follow.

Let $\Omega \subset \mathbb{C}^n$ be a domain, containing the origin. We call a holomorphic mapping $f(z) = (f_1(z), \ldots, f_n(z))$ from Ω into $\mathbb{C}^n$ *normalized* if $f(0) = 0$, $J_f(0) = I$, the identity matrix, where $z = (z_1, \ldots, z_n)$.

In 1988, Barnard, FitzGerald, and Gong [BFG1] proved the following result. Let $f(z) = (f_1(z), f_2(z))$, where $z = (z_1, z_2) \in B^2$, and

$$\begin{cases} f_1(z) = z_1 + d_{2,0}^1 z_1^2 + d_{1,1}^1 z_1 z_2 + d_{0,2}^1 z_2^2 + \cdots, \\ f_2(z) = z_2 + d_{2,0}^2 z_1^2 + d_{1,1}^2 z_1 z_2 + d_{0,2}^2 z_2^2 + \cdots, \end{cases}$$

be a holomorphic mapping in a family $\mathcal{S}$ of $\mathcal{A}$-invariant of normalized holomorphic mappings on B^2. Then

$$\left| \log \left[(1 - z\bar{z}')^{3/2} \det J_f(z) \right] \right| \leqslant C(\mathcal{S}) \log \frac{1 + |z|}{1 - |z|},$$

where $C(\mathcal{S}) = \sup\{|d_{2,0}^1 + d_{1,1}^2| \mid f \in \mathcal{S}\}$.

From this inequality, it follows immediately the following distortion theorem:

$$\frac{(1-|z|)^{C(\mathcal{S})-3/2}}{(1+|z|)^{C(\mathcal{S})+3/2}} \leqslant |\det J_f(z)| \leqslant \frac{(1+|z|)^{C(\mathcal{S})-3/2}}{(1-|z|)^{C(\mathcal{S})+3/2}},$$

and

$$\left| \arg \det J_f(z) \right| \leqslant C(\mathcal{S}) \log \frac{1+|z|}{1-|z|}.$$

In particular, if $\mathcal{S}$ is the family of normalized biholomorphic convex mapping, then one can prove $1.5 \leqslant C(\mathcal{S}) \leqslant 1.761$. And it is conjectured that $C(\mathcal{S}) = 3/2$. The normalized biholomorphic convex mapping $\left(\dfrac{z_1}{1-z_1}, \dfrac{z_2}{1-z_1} \right)$ is one mapping which attains this value.

We note that if the conjecture is true, then the following inequality:

$$(1+|z|)^{-3} \leqslant |\det J_f(z)| \leqslant (1-|z|)^{-3}$$

holds.

Taishun Liu [**LiuT1**] extended the previous result to the unit ball B^n in $\mathbb{C}^n$. Let $f(z) = (f_1(z), \ldots, f_n(z)) \in \mathcal{S}$, where $\mathcal{S}$ is a family of $\mathcal{A}$-invariant normalized holomorphic mappings on B^n and $z = (z_1, \ldots, z_n) \in \mathbb{C}^n$. If $f(z)$ can be expressed as

$$f(z) = z + (zA^1 z', \ldots, zA^n z') + \cdots$$

where $A^i = (a^i_{j,k})_{1 \leqslant j,k \leqslant n}$, $i = 1, \ldots, n$; then

$$\left| \log \left[(1 - z\bar{z}')^{(n+1)/2} \det J_f(z) \right] \right| \leqslant C(\mathcal{S}) \log \frac{1+|z|}{1-|z|},$$

where $C(\mathcal{S}) = \sup \left\{ |\sum_{i=1}^{n} a^i_{i1}| \mid f \in \mathcal{S} \right\}$. He also proved that if $\mathcal{S}$ is the family of normalized biholomorphic convex mappings, then $\dfrac{n+1}{2} \leqslant C(\mathcal{S}) \leqslant 1 + \dfrac{\sqrt{2}(n-1)}{2}$.

It is conjectured by Barnard, FitzGerald and Gong [**BFG1**] that $C(\mathcal{S}) = \dfrac{n+1}{2}$ if $\mathcal{S}$ is the family of normalized biholomorphic convex mappings. The biholomorphic convex mapping $\left(\dfrac{z_1}{1-z_1}, \ldots, \dfrac{z_n}{1-z_1} \right)$ is one of the mappings which attains this value. Furthermore, if the conjecture is true, then one can derive

$$(1+|z|)^{-n-1} \leqslant |\det J_f(z)| \leqslant (1-|z|)^{-n-1}.$$

However, we do not know whether the conjecture is true or not.

2. Jacobian determinant of the biholomorphic mappings on transitive domains.

Let $M \subset \mathbb{C}^n$ be a transitive domain, bounded or unbounded, m be a point in M. Let G be a Lie group consisting of some holomorphic automorphisms of M and acting transitively on M, K be an isotropy group of G which leaves m fixed. We denote by ψ_z a holomorphic automorphism of M which maps

$z \in M$ to $m \in M$, i.e., $\psi_z(z) = m$ and denote the Jacobian of ψ_z by J_{ψ_z}. If M is unbounded, we must assume $|\det J_{\psi_k}(m)| = 1$ for all $k \in K$. Set $K(z, \bar{z}) = c \det J_{\psi_z}(z)\overline{\det J_{\psi_z}(z)}$ with c being a constant and denote

$$K(m, \bar{m})^{-1} \left(\frac{\partial}{\partial z_p} K(z, \bar{z}) \right)_{z=m}$$

by C_p, then $K(z, \bar{z})$ is well defined and if M is bounded, $K(z, \bar{z})$ is the Bergman kernel function for certain constant c.

Suppose f is a biholomorphic mapping of M into $\mathbb{C}^n$. Then $\det J_f(z)$ can be expressed in terms of $K(z, \bar{z})$, C_p, and the coefficients of the expansion of f. This is the result of Gong and Zheng [**GZ1**].

Under the same assumptions given above, we denote by ϕ_z the inverse of ψ_z. For any z in M, we can join z and m by an analytic curve $\sigma(\rho) = (\sigma_1(\rho), \ldots, \sigma_n(\rho))$ $(0 \leqslant \rho \leqslant 1)$ such that $\sigma(0) = m$ and $\sigma(1) = z$. Let $a = (a_1, \ldots, a_n) = \sigma(\rho)$, $F(w) = F_a(w) = (f(\phi_a(w)) - f(z))J_f(a)^{-1}J_{\phi_a}(m)^{-1}$,

$$d_{ij}(\rho) = \left(d_{ij}^{(1)}(\rho), \ldots, d_{ij}^{(n)}(\rho) \right) = \frac{1}{2} \frac{\partial F(w)}{\partial w_i \partial w_j}\bigg|_{w=m} \qquad 1 \leqslant i, j \leqslant n,$$

$J_{\phi_a}(a) = (u_{p\ell})$ and $J_f(z)$ be the Jacobian of f at z. Then the following explicit formula for the determinant of the Jacobian can be derived:

$$\log \det J_f(z) - \log \det J_f(m) = \frac{1}{2}(\log K(z, \bar{z}) - \log K(m, \bar{m}))$$

$$+ \sum_{\ell, k=1}^{n} \int_0^1 \frac{da_k}{d\rho} u_{k\ell}(\rho) \left(2 \sum_{j=1}^{n} d_{ij}^{(j)}(\rho) - c_l \right) d\rho$$

$$+ i \operatorname{Im} \int_0^1 \sum_{j=1}^{n} \frac{da_j}{d\rho} \frac{\partial}{\partial a_j} K(a, \bar{a}) K(a, \bar{a})^{-1} d\rho.$$

Moreover, if M is starlike with respect to m, then

$$\log \det J_f(z) - \log \det J_f(m) = \frac{1}{2}(\log K(z, \bar{z}) - \log K(m, \bar{m}))$$

$$+ \sum_{\ell, k=1}^{n} \int_0^1 (z_k - m_k) u_{k\ell}(\rho) \left(2 \sum_{j=1}^{n} d_{ij}^{(j)}(\rho) - c_l \right) d\rho$$

$$+ i \operatorname{Im}\left\{ \int_0^1 K(a, \bar{a})^{-1} \sum_{k=1}^{n} (z_k - m_k) \frac{\partial}{\partial a_k} K(a, \bar{a}) d\rho \right\}.$$

Additionally, if M is bounded and circular, then

$$\log \det J_f(z) - \log \det J_f(m) = \frac{1}{2}(\log K(z, \bar{z}) - \log K(m, \bar{m}))$$

$$+ 2 \sum_{j,k,\ell} \int_0^1 (z_k - m_k) u_{k\ell}(\rho) d_{ij}^{(j)}(\rho) \, d\rho$$

$$+ i \operatorname{Im} \left\{ \int_0^1 K(a, \bar{a})^{-1} \sum_{k=1}^n (z_k - m_k) \frac{\partial}{\partial a_k} K(a, \bar{a}) \, d\rho \right\}.$$

All these formulas will enable one to get some estimations of the determinant of the Jacobian.

3. Distortion theorem for the family of $\mathcal{A}$-invariant mappings on bounded symmetric domains.

Let $M \subset \mathbb{C}^n$ be a bounded symmetric domain, containing the origin, and M be the canonical Harish-Chandra realization of the Hermitian symmetric space G/K, where G is the group of holomorphic automorphisms of M, and K is the isotropy subgroup in G which leave the origin fixed. Let $\mathfrak{g}$ be the Lie algebra of G, $\mathfrak{k}$ be the maximal compact subalgebra of $\mathfrak{g}$ which corresponds to K. Then $\mathfrak{g}$ has the Cartan decomposition $\mathfrak{g} = \mathfrak{k} + \mathfrak{p}$. If $\mathfrak{a}$ is the maximal Abelian subspace in $\mathfrak{p}$, then we can choose $X_1, \ldots, X_q$ a basis of $\mathfrak{a}$, where $q = \dim \mathfrak{a} = \operatorname{rank} G/K$, and for any $X \in \mathfrak{a}$, there exists a unique decomposition $X = x_1 X_1 + \cdots + x_q X_q$. If A is the analytic subgroup in G corresponding to $\mathfrak{a}$ in $\mathfrak{g}$, then G has Iwasawa decomposition $G = KAN$. For every $z \in M$, there exist $X \in \mathfrak{a}$ and $k \in K$, such that $z = \xi(\exp \operatorname{Ad}(k) X \cdot O)$, where O is the identity coset in G/K, ξ is the canonical holomorphic diffeomorphism of G/K onto M. Because K acts on M just as a subgroup of the unitary group $U(n)$ acts on $\mathbb{C}^n$, we have

$$z = \xi(\exp \operatorname{Ad}(k) X \cdot O) = k \xi(\exp X \cdot O) = \Lambda \tilde{k},$$

where $\Lambda = \xi(\exp X \cdot O) = (\tanh x_1, \ldots, \tanh x_q, 0, \ldots, 0) \in M$, $\tilde{k} = (k_{ij}) \in U_n$.

We say that the holomorphic mapping $f(z) = (f_1(z), \ldots, f_n(z))$ which maps M into $\mathbb{C}^n$ is normalized if

$$f(z) = z + \sum_{i,j}^n d_{ij} z_i z_j + \cdots,$$

where $d_{ij} = \left(d_{ij}^{(1)}, \ldots, d_{ij}^{(n)} \right)$, $z = (z_1, \ldots, z_n)$. A family $\mathcal{S}_M$ of normalized holomorphic mappings of M into $\mathbb{C}^n$ is called an $\mathcal{A}$-invariant family if the following condition is satisfied: the composition of any $f(z) \in \mathcal{S}_M$ with any holomorphic automorphism of M, after normalization, remains a holomorphic mapping in $\mathcal{S}_M$.

Gong and Zheng [GZ1] proved the following result.

Under the same assumptions given above, if f is a normalized biholomorphic mapping of M into $\mathbb{C}^n$, $f \in \mathcal{S}_M$, where $\mathcal{S}_M$ is a family of $\mathcal{A}$-invariant biholomorphic mappings, $z = \xi(\exp \operatorname{Ad}(k) X \cdot O) \in M$, $X = x_1 X_1 + \cdots + x_q X_q$,

$q = \dim M = \operatorname{rank} G/K$, then the following inequality

$$\left| \log \frac{\det J_f(z)}{(K(z,\bar{z})/K(0,0))^{1/2}} \right| \leqslant C(\mathcal{S}_M) \sum_{p=1}^{q} \log \frac{1 + |\tanh x_p|}{1 - |\tanh x_p|},$$

holds, where $C(\mathcal{S}_M) = \sup \left\{ \left| \sum_{i,j} k_{\ell i} d_{ij}^{(j)} \right| \mid f \in \mathcal{S}_M, \ell = 1, \ldots, q \right\}$, $(k_{\ell i})_{1 \leqslant i, \ell \leqslant n} \in U_n$, and K_M is the Bergman kernel function of M.

This inequality implies the following distortion theorem:

$$\sqrt{\frac{K(z,\bar{z})}{K(0,0)}} \left(\prod_{p=1}^{q} \frac{1 - |\tanh x_p|}{1 + |\tanh x_p|} \right)^{C(\mathcal{S}_M)} \leqslant |\det J_f(z)|$$

$$\leqslant \sqrt{\frac{K(z,\bar{z})}{K(0,0)}} \left(\prod_{p=1}^{q} \frac{1 + |\tanh x_p|}{1 - |\tanh x_p|} \right)^{C(\mathcal{S}_M)}.$$

A more precise estimation is established in Gong-Zhang [**GZ2**] as follows:

$$\sqrt{\frac{K(z,\bar{z})}{K(0,0)}} \prod_{p=1}^{q} \left(\frac{1 - |\tanh x_p|}{1 + |\tanh x_p|} \right)^{C_p(\mathcal{S}_M)} \leqslant |\det J_f(z)|$$

$$\leqslant \sqrt{\frac{K(z,\bar{z})}{K(0,0)}} \prod_{p=1}^{q} \left(\frac{1 + |\tanh x_p|}{1 - |\tanh x_p|} \right)^{C_p(\mathcal{S}_M)}.$$

4. Distortion theorem for the family of biholomorphic starlike mappings on bounded symmetric domains.

With the same assumptions given in the previous section on M, the family of biholomorphic convex mappings on M is a family of biholomorphic $\mathcal{A}$-invariant mappings, which will be discussed in section IV. Note that the family of biholomorphic starlike mappings is not a family of biholomorphic $\mathcal{A}$-invariant mappings. Thus we can not make use of the results in the previous section. Instead the following result is obtained by Gong-Zheng [**GZ4**]:

Suppose the assumptions on M are same as those stated in previous section. Let $\mathcal{S}(M)$ denote the family of normalized biholomorphic starlike mappings from M into $\mathbb{C}^n$. Then there exists a positive constant $C(\mathcal{S}(M))$, such that for any $f(z) \in \mathcal{S}(M)$, the following inequality

$$\sqrt{\frac{K(z,\bar{z})}{K(0,0)}} \liminf_{\substack{b \in \partial M \\ \theta \to b}} \left(\frac{K(\theta, \bar{\theta})}{K(\eta, \bar{\eta})} \right)^{C(\mathcal{S}(M))} \leqslant |\det J_f(z)|$$

$$\leqslant \sqrt{\frac{K(z,\bar{z})}{K(0,0)}} \limsup_{\substack{b \in \partial M \\ \theta \to b}} \left(\frac{K(\theta, \bar{\theta})}{K(\eta, \bar{\eta})} \right)^{C(\mathcal{S}(M))}.$$

holds, where K is the Bergman kernel function of M, $z = \xi(ka \cdot O)$, $\theta = (a_1 \cdot O)$, $\eta = \xi(a_1 a \cdot O)$, $k \in K$, $a, a_1 \in A$. Furthermore $C(\mathcal{S}(M))$ is bounded by

$$\frac{1}{2} \leqslant C(\mathcal{S}(M)) \leqslant \frac{7n - 2}{4},$$

and it is conjectured that $C(\mathcal{S}(M)) = 1$.

For the family of biholomorphic convex mappings $\mathcal{C}(M)$, a similar result holds. There exists a positive constant $C(\mathcal{C}(M))$, such that for any $f(z) \in \mathcal{C}(M)$, the following inequality

$$\sqrt{\frac{K(z,\bar{z})}{K(0,0)}} \liminf_{\substack{b \in \partial M \\ \theta \to b}} \left(\frac{K(\theta,\bar{\theta})}{K(\eta,\bar{\eta})}\right)^{C(\mathcal{C}(M))} \leqslant |\det J_f(z)|$$

$$\leqslant \sqrt{\frac{K(z,\bar{z})}{K(0,0)}} \limsup_{\substack{b \in \partial M \\ \theta \to b}} \left(\frac{K(\theta,\bar{\theta})}{K(\eta,\bar{\eta})}\right)^{C(\mathcal{C}(M))}$$

holds, the meanings of K, θ and η are as before. Moreover $C(\mathcal{C}(M))$ is bounded by

$$\frac{1}{2} \leqslant C(\mathcal{C}(M)) \leqslant \frac{2n-1}{2}$$

and it is conjectured that $C(\mathcal{C}(M)) = \frac{1}{2}$.

III. Growth theorem and covering theorem for biholomorphic starlike mappings

1. Growth theorem and covering theorem for biholomorphic starlike mappings on the unit ball.

H. Cartan [**CarH**] cited the growth theorem of univalent function of one complex variable as one of the results that could not be extended to the polydisc (nor to the ball) of several complex variables. He also observed that for normalized biholomorphic mappings there is no neighborhood about the origin that is always covered by the range of a biholomorphic mapping of the polydisc (nor the ball). That is, there is no Koebe $\frac{1}{4}$-theorem for these mappings. In 1988, Barnard, FitzGerald and Gong [**BFG2**] gave a growth theorem and a covering theorem for the normalized biholomorphic starlike mapping on the unit ball B^n and derived the following growth theorem.

Let $f(z) = (f_1(z), \ldots, f_n(z))$ be a normalized biholomorphic starlike mapping of the unit ball B^n in $\mathbb{C}^n$. Then for any point $z = (z_1, \ldots, z_n) \in B^n$, the inequality

$$\frac{|z|}{(1+|z|)^2} \leqslant |f(z)| \leqslant \frac{|z|}{(1-|z|)^2}$$

holds, where $|z| = (|z_1|^2 + \cdots + |z_n|^n)^{1/2}$. The estimates are sharp.

As a consequence of this result, the following covering theorem is derived.

If $f(z)$ is a normalized biholomorphic starlike mapping on the unit ball B^n in $\mathbb{C}^n$, then the image of f contains a ball of radius $\frac{1}{4}$ centered at the origin. The value $\frac{1}{4}$ is the best possible.

There are many mappings that make the inequalities sharp in the growth theorem. For example, $f(z) = \left(\dfrac{z_1}{(1-z_1)^2}, \dfrac{z_2}{(1-z_2)^2}\right)$ is a normalized biholomorphic starlike mapping on the unit ball in $\mathbb{C}^2$, which makes the equalities in the growth theorem hold. Moreover, consider the mappings $f(z) = \left(\dfrac{z_1}{(1-z_1)^2}, f_2(z_2) \right)$ where $f_2(z_2)$ is any normalized univalent analytic function of one variable z_2 that maps the unit disc onto a domain which is starlike with respect to the origin. These mappings are normalized biholomorphic starlike mappings that enables the equalities in the growth theorem to hold for certain values of z.

It is easy to see that the estimate in the covering theorem is sharp also. The examples just given show that the point $w_1 = -\frac{1}{4}$, $w_2 = 0$ can fail to be in the range.

If k-fold symmetry of f is assumed, that is, the image of f is unchanged when multiplied by the scalar complex number $\exp(2\pi i/k)$ for some positive integer k, then the growth theorem and the covering theorem can be strengthened as follows.

If f is normalized biholomorphic starlike mapping in the unit ball in $\mathbb{C}^n$ with a k-fold symmetric image, then

$$\frac{|z|}{(1+|z|^k)^{2/k}} \leqslant |f(z)| \leqslant \frac{|z|}{(1-|z|^k)^{2/k}}$$

for any $z \in B^n$. The image of the ball under f contains a ball of radius $2^{-2/k}$. All these estimates are the best possible.

The previous result shows that not all balanced domains can be mapped onto each other by biholomorphic mappings. (A domain D in $\mathbb{C}^n$ is said to be balanced if $\xi w \in D$ whenever $w \in D$ and $\xi \in \mathbb{C}$ with $|\xi| \leqslant 1$.) Specifically the following result is obtained.

The only balanced domain that is the image of the unit ball under a normalized biholomorphic mapping is the unit ball.

As a corollary of this work, we demonstrate the well known Poincare theorem: there is no biholomorphic mapping of the unit ball in $\mathbb{C}^n$ into the unit polydisc in $\mathbb{C}^n$ in case $n \geqslant 2$.

2. Growth theorem and covering theorem for biholomorphic starlike mappings on Reinhardt domain B_p.

Gong, Wang and Yu [**GWY1**] extended the results of the previous section to the case of a class of Reinhardt domain.

Let $B_p = \left\{ z = (z_1, \ldots, z_n) \in \mathbb{C}^n \mid \sum_{j=1}^n |z_j|^p < 1 \right\}$ for any $p > 1$, which is a class of Reinhardt domains. When $p = 2$, B_2 is the unit ball in $\mathbb{C}^n$. Let $f(z) = (f_1(z), \ldots, f_n(z))$, $z = (z_1, \ldots, z_n) \in \mathbb{C}^n$, be a normalized biholomorphic starlike mapping on B_p which maps B_p into $\mathbb{C}^n$. Then

$$\frac{|z|}{(1+\|z\|_p)^2} \leqslant |f(z)| \leqslant \frac{|z|}{(1-\|z\|_p)^2}, \qquad z \in B_p,$$

where $\|z\|_p = \left(\sum_{i=1}^n |z_i|^p \right)^{1/p}$. The estimates are precise. The examples in the previous section are normalized biholomorphic starlike on B_p, which can be verified with the use of the criterion of Suffridge [**S1**] (cf. section I.1). Also, these examples will make the equalities in the growth theorem attainable. As a consequences of the growth theorem, the following estimates hold:

$$n^{(p-2)/2p}\frac{\|z\|_p}{(1+\|z\|_p)^2} \leqslant |f(z)| \leqslant \frac{\|z\|_p}{(1-\|z\|_p)^2}, \qquad \text{if } 2 > p > 1;$$

$$\frac{\|z\|_p}{(1+\|z\|_p)^2} \leqslant |f(z)| \leqslant n^{(p-2)/2p}\frac{\|z\|_p}{(1-\|z\|_p)^2}, \qquad \text{if } p \geqslant 2.$$

Thus, it follows the covering theorem: $f(B_p)$ contains a ball of radius $1/4k$ centered at the origin, where $k = n^{(2-p)/2p}$ if $1 < p < 2$; $k = 1$ if $p \geqslant 2$.

In section I.4, we introduced the Loewner chain in several complex variables, which was established by J. A. Pfaltzgraff [**Pfa1**]. Using it, Pfaltzgraff proved another kind of growth theorem on B_p: If $f(z)$ is the normalized biholomorphic starlike mappings of B_p into $\mathbb{C}^n$, then the following inequality

$$\frac{\|z\|_p}{(1+\|z\|_p)^2} \leqslant \|f(z)\|_p \leqslant \frac{\|z\|_p}{(1-\|z\|_p)^2}, \qquad z \in B_p$$

holds. The estimate is precise. As a corollary, it implies that $f(B_p)$ contains $\frac{1}{4}B_p$, and that the constant $\frac{1}{4}$ is best possible. In another paper by Gong, Wang and Yu [**GWY2**], the authors proved that these two kinds of growth theorems are equivalent to each other, i.e., one growth theorem implies the other.

Thus far, no growth theorem and the covering theorem for the normalized biholomorphic starlike mappings on D_p has been found yet (cf. section I.1).

3. Growth theorem and covering theorem for biholomorphic starlike mappings on classical domains.

The growth theorem and the covering theorem on the unit ball can be extended to four classical domains. Taishun Liu [**LiuT2**] proved the following growth theorem and covering theorem.

If $f(z)$ is a normalized biholomorphic starlike mapping of classical domain $\mathcal{R}_K$ into $\mathbb{C}^{\dim \mathcal{R}_K}$ ($K = I, II, III, IV$), then for any $Z \in \mathcal{R}_K$,

$$\frac{\|Z\|_K}{(1+\|Z\|_K)^2} \leqslant \|f(Z)\|_K \leqslant \frac{\|Z\|_K}{(1-\|Z\|_K)^2},$$

where $\|\cdot\|_K$ is the norm of the element of $\mathcal{R}_K$ (cf. section I.2). The estimates are precise, i.e., there exist mappings that make the equalities attainable. For instance, under the mappings $f(Z) = (1-z_{11})^{-2}Z$ in the case of $\mathcal{R}_I$; $f(Z) = (1-z_{11})^{-2}Z$, $Z = Z'$ in the case of $\mathcal{R}_{II}$; $f(Z) = (1-z_{12})^{-2}Z$, $Z = -Z'$ in the case of $\mathcal{R}_{III}$ and $f(z) = (1-z_1)^{-2}z$ in the case of $\mathcal{R}_{IV}$; make the equalities attainable.

As a consequence, we get the Koebe $\frac{1}{4}$-covering theorem: If $f(Z)$ is a normalized biholomorphic starlike mapping of $\mathcal{R}_K$ which maps $\mathcal{R}_K$ into $\mathbb{C}^{\dim \mathcal{R}_K}$ ($K = I, II, III, IV$), then $f(Z)$ contains $\frac{1}{4}\mathcal{R}_K$, the constant $\frac{1}{4}$ being the best possible. All the previous examples attain the constant $\frac{1}{4}$.

4. Another growth theorem for biholomorphic starlike mappings on classical domains.

In the last section we stated a growth theorem of normalized biholomorphic starlike mappings on classical domains, which estimates the norm of $f(Z)$ in terms of the norm of Z. Consequently, we get the corresponding covering theorems of these mappings. In this section we give an another growth theorem, which is the estimations of the modulus of $f(Z)$ by the modulus and the norm of z. These are results of Gong, Wang and Yu [**GWY2**]. In that same paper, it is shown that these two growth theorems are equivalent to each other.

Let $f(Z)$ be a normalized biholomorphic starlike mapping on $\mathcal{R}_K$ ($K = I, II, III, IV$), which maps $\mathcal{R}_K$ into $\mathbb{C}^{\dim \mathcal{R}_K}$. Then for any $Z \in \mathcal{R}_K$,

$$\frac{|Z|}{(1+\|Z\|_K)^2} \leqslant |f(Z)| \leqslant \frac{|Z|}{(1-\|Z\|_K)^2},$$

where $\|\cdot\|_K$ is the norm of the element of $\mathcal{R}_K$, $|Z| = \sqrt{\operatorname{tr}(Z\bar{Z}')}$ and $|f(Z)| = \sqrt{\operatorname{tr}(f(Z)\overline{f(Z)}')}$. The estimates are precise. For instance, the equalities hold if the mappings are the examples in the previous section. Taishun Liu [**LiuT2**] derived this growth theorem in the case of $\mathcal{R}_{IV}$.

From this growth theorem, we get another kind of Koebe covering theorem. If f is defined as above, then $f(\mathcal{R}_I) \supset \frac{1}{4}B^{mn}$, $f(\mathcal{R}_{II}) \supset \frac{1}{4}B^{n^2} \cap V_{II}$, where $V_{II} = \{Z \in \mathbb{C}^{n^2} \mid Z' = Z\}$, $f(\mathcal{R}_{III}) \supset \frac{\sqrt{2}}{4}B^{n^2} \cap V_{III}$, where $V_{III} = \{Z \in \mathbb{C}^{n^2} \mid Z' = -Z\}$, $f(\mathcal{R}_{IV}) \supset \frac{\sqrt{2}}{8}B^n$. All these constants are precise. The examples in previous section are extremal mappings.

5. Growth theorem and covering theorem for biholomorphic starlike mappings on Banach space.

Let X and Y be Banach spaces and $B_r = \{x \in X \mid \|x\| < r\}$, $B = B_1$. Suppose that U is an open subset of X and $f : U \to Y$ then f is holomorphic in U if given $x \in U$, there is a bounded linear map $Df(x) : X \to Y$ such that

$$\lim_{h \to 0} \frac{\|f(x+h) - f(x) - Df(x)(h)\|}{\|h\|} = 0.$$

The linear map $Df(x)$ is called the Frechet derivative of f at x. If f is holomorphic in U and $x \in U$, then for $n = 1, 2, \ldots$, there is a bounded symmetric n-linear map $D^n f(x) : X \times X \times \cdots \times X \to Y$ such that $f(y) = \sum_{n=0}^{\infty} \frac{1}{n!} D^n f(x)((y-x)^n)$ for all y in some neighborhood of x. Here $D^0 f(x)((y-x)^0) = f(x)$ and for $n \geqslant 1$, $D^n f(x)((y-x)^n) = D^n f(x)(y-x, \ldots, y-x)$. A mapping $f : U \to Y$ is biholomorphic if the inverse mapping f^{-1} exists and is holomorphic on an open set $V \subset Y$ and $f^{-1}(V) = U$. A mapping f is locally biholomorphic if given $y \in f(U)$ there is a neighborhood V of y such that f^{-1} exists and is holomorphic in V. $Df(x)$ has a bounded inverse for each $x \in U$ if and only if f is locally biholomorphic.

Let X^* be the dual of X and given $x \in X$, $x \neq 0$. Define

$$T(x) = \{x^* \in X^* \mid x^*(x) = \|x\| \text{ and } \|x^*\| = 1\}.$$

Let N denote the set of all holomorphic mappings $w : B \to X$ such that $w(0) = 0$ and $\operatorname{Re} x^*(w(x)) > 0$ when $x \in B$, $x \neq 0$ and $x^* \in T(x)$. Let $M = \{w \in N \mid Dw(0) = 1\}$.

Suffridge [**S2**] proved: Suppose $f : B \to Y$ is locally biholomorphic with $f(0) = 0$. Then f is starlike if and only if $f(x) = Df(x)(w(x))$ where $w \in M$. He also proved: If $f : B \to Y$ is a biholomorphic convex mapping, then

$$D^2 f(x)(x, x) + Df(x)(x) = Df(x)(w(x)),$$

where $w \in M$. Suppose $f : B \to Y$ is locally biholomorphic and set $f(x) - f(y) = Df(x)(w(x, y))$, $x, y \in B$. Then f is convex if and only if $\operatorname{Re} x^*(w(x, y)) > 0$ whenever $\|y\| < \|x\|$ and $x^* \in T(x)$. He also defined the close-to-convex mappings and the close-to-starlike mappings and gave numerous interesting examples (cf. Suffridge [**S3**]).

Based upon the results of Suffridge, Dong and Zhang [**DZ**] proved the following growth theorem of the unit ball on Banach space.

Let B be the unit ball of Banach space and f be a normalized biholomorphic starlike mapping on B (*normalized* means $f(0) = 0$, $Df(0) = I$). Then for any $x \in B$, the inequality

$$\frac{\|x\|}{(1 + \|x\|)^2} \leqslant \|f(x)\| \leqslant \frac{\|x\|}{(1 - \|x\|)^2}$$

holds. These estimates are precise for some special Banach spaces.

As a consequence, one can obtain the covering theorem of the Banach space as follows.

Let B be the unit ball of Banach space X and f be the normalized biholomorphic starlike map on B. Then $f(B)$ contains $\frac{1}{4}B$. This estimate is precise for some special Banach spaces.

Of course, we can treat the unit ball, the Reinhardt domain B_p ($p > 1$) and the classical domains of finite dimension complex space as the unit ball in Banach spaces if we define some norm on these domains first. Then, the growth theorem of the normalized biholomorphic starlike mappings on the unit ball in Banach space is another extension of some results of sections III. 1, III. 2 and III. 3.

IV. Biholomorphic convex mappings.

1. Distortion theorem for biholomorphic convex mappings on bounded symmetric domains.

Let Ω be an irreducible bounded symmetric domain of rank $\geqslant 2$. We have the following distortion theorems. (cf. Gong and Zheng [**GZ2, GZ3**])

For $\mathcal{R}_I : I - Z\bar{Z}' > 0$, $Z = Z^{(m,n)} \in \mathbb{C}^{m \times n}$ ($2 \leqslant m \leqslant n$), Z can be express as

$$Z = U \begin{pmatrix} \lambda_1 & 0 & \cdots & 0 & 0 & \cdots & 0 \\ 0 & \lambda_2 & \cdots & 0 & 0 & \cdots & 0 \\ \vdots & \vdots & \ddots & \vdots & \vdots & & \vdots \\ 0 & 0 & \cdots & \lambda_m & 0 & \cdots & 0 \end{pmatrix} V,$$

where $U \in U_m$, $V \in U_n$, $1 > \lambda_1 \geqslant \cdots \geqslant \lambda_m \geqslant 0$. Let $f(Z) : \mathcal{R}_I \to \mathbb{C}^{m \times n}$ be a normalized biholomorphic convex mapping. Then

$$\prod_{j=1}^m \frac{(1-\lambda_j)^{C(\mathcal{S})-(m+n)/2}}{(1+\lambda_j)^{C(\mathcal{S})+(m+n)/2}} \leqslant |\det J_j(Z)| \leqslant \prod_{j=1}^m \frac{(1+\lambda_j)^{C(\mathcal{S})-(m+n)/2}}{(1-\lambda_j)^{C(\mathcal{S})+(m+n)/2}},$$

where $\frac{1}{2}(m+n) \leqslant C(\mathcal{S}) \leqslant \frac{1}{2}(mn+1) + \frac{\sqrt{2}-1}{2}(m+n-2)$. We conjecture that

$$\prod_{j=1}^m (1+\lambda_j)^{-m-n} \leqslant |\det J_f(Z)| \leqslant \prod_{j=1}^m (1-\lambda_j)^{-m-n},$$

and the estimate is sharp.

For $\mathcal{R}_{II} : I - Z\bar{Z} > 0$, $Z = Z' = Z^{(n)} \in \mathbb{C}^{\frac{1}{2}n(n+1)}$ ($2 \leqslant n$), Z can be expressed as

$$Z = U \begin{pmatrix} \lambda_1 & & 0 \\ & \ddots & \\ 0 & & \lambda_n \end{pmatrix} U',$$

where $U \in U_n$, $1 > \lambda_1 \geqslant \cdots \geqslant \lambda_n \geqslant 0$. Let $f(Z) : \mathcal{R}_{II} \to \mathbb{C}^{\frac{1}{2}n(n+1)}$ be a normalized biholomorphic convex mapping. Then

$$\prod_{j=1}^n \frac{(1-\lambda_j)^{C(\mathcal{S})-(n+1)/2}}{(1+\lambda_j)^{C(\mathcal{S})+(n+1)/2}} \leqslant |\det J_f(Z)| \leqslant \prod_{j=1}^n \frac{(1+\lambda_j)^{C(\mathcal{S})-(n+1)/2}}{(1-\lambda_j)^{C(\mathcal{S})+(n+1)/2}},$$

where $\frac{1}{2}(n+1) \leqslant C(\mathcal{S}) \leqslant \frac{1}{2}(\frac{1}{2}n(n+1)+1) + (\sqrt{3}-1)(n-1)$. We conjecture that

$$\prod_{j=1}^n (1+\lambda_j)^{-n-1} \leqslant |\det J_f(Z)| \leqslant \prod_{j=1}^n (1-\lambda_j)^{-n-1}$$

and the estimate is sharp.

For $\mathcal{R}_{III} : I + Z\bar{Z} > 0$, $Z = -Z' = Z^{(n)} \in \mathbb{C}^{\frac{1}{2}n(n-1)}$ ($3 \leqslant n$), Z can be expressed as

$$Z = UMU', \qquad U \in U_n,$$

where

$$M = \begin{bmatrix} 0 & \lambda_1 \\ -\lambda_1 & 0 \end{bmatrix} + \cdots,$$

$1 > \lambda_1 \geqslant \lambda_2 \geqslant \cdots \geqslant \lambda_\nu \geqslant 0$, $\nu = [\frac{1}{2}n]$, the direct sum of M end at $\begin{bmatrix} 0 & \lambda_\nu \\ -\lambda_\nu & 0 \end{bmatrix}$ if n is even, and at 0 if n is odd. Let $f(Z) : \mathcal{R}_{III} \to \mathbb{C}^{\frac{1}{2}n(n-1)}$ be a normalized biholomorphic convex mapping. Then

$$\prod_{j=1}^{\nu} \frac{(1-\lambda_j)^{C(\mathcal{S})-n+1}}{(1+\lambda_j)^{C(\mathcal{S})+n-1}} \leqslant |\det J_f(Z)| \leqslant \prod_{j=1}^{\nu} \frac{(1+\lambda_j)^{C(\mathcal{S})-n+1}}{(1-\lambda_j)^{C(\mathcal{S})+n-1}}$$

where $n-1 \leqslant C(\mathcal{S}) \leqslant \frac{1}{2}\left(\frac{1}{2}n(n-1)+1\right) + (\sqrt{2}-1)(n-2)$. We conjecture that

$$\prod_{j=1}^{v}(1+\lambda_j)^{-2n+2} \leqslant |\det J_f(Z)| \leqslant \prod_{j=1}^{v}(1-\lambda_j)^{-2n+2}$$

and the estimate is sharp.

For $\mathcal{R}_{IV} : 1 - 2\bar{z}z' + |zz'|^2 > 0$, $|zz'| < 1$, $z \in \mathbb{C}^n$ $(2 \leqslant n)$, z can be expressed as

$$z = e^{i\theta}\left(\frac{1}{2}(\lambda_1 + \lambda_2), \frac{i}{2}(\lambda_1 - \lambda_2), 0, \ldots, 0\right)\Gamma,$$

$\Gamma \in SO(n)$, $1 > \lambda_1 \geqslant \lambda_2 \geqslant 0$. Let $f(z) : \mathcal{R}_{IV} \to \mathbb{C}^n$ be a normalized biholomorphic convex mapping. Then

$$\frac{((1-\lambda_1)(1-\lambda_2))^{C(\mathcal{S})-n/2}}{((1+\lambda_1)(1+\lambda_2))^{C(\mathcal{S})+n/2}} \leqslant |\det J_f(z)| \leqslant \frac{((1+\lambda_j)(1+\lambda_j))^{C(\mathcal{S})-n/2}}{((1-\lambda_j)(1-\lambda_j))^{C(\mathcal{S})+n/2}}$$

where $\frac{1}{2}n \leqslant C(\mathcal{S}) \leqslant n - \frac{1}{2}$. We conjecture that

$$\prod_{j=1}^{2}(1+\lambda_j)^{-n} \leqslant |\det J_f(z)| \leqslant \prod_{j=1}^{2}(1-\lambda_j)^{-n}$$

and the estimate is sharp.

2. Growth theorem and distortion theorem for biholomorphic convex mappings on the unit ball and on the Reinhardt domains..

T. J. Suffridge [S4], C. Thomas [T] and Taishun Liu [LiuT2] independently used different methods to prove the growth theorem for normalized biholomorphic convex mappings on the unit ball in $\mathbb{C}^n$.

Let $f(z) = (f_1(z), \ldots, f_n(z))$ be a normalized biholomorphic convex mapping on the unit ball $B^n : z\bar{z}' < 1$ in $\mathbb{C}^n$. Then for any point $z \in B^n$,

$$\frac{|z|}{1+|z|} \leqslant |f(z)| \leqslant \frac{|z|}{1-|z|}.$$

The estimate is sharp. There are many mappings that will make the equalities attainable.

As a consequence, we have the covering theorem. Let $f(z)$ be a normalized biholomorphic convex mapping on the unit ball B^n. Then the image of the mapping contains a ball centered at the origin with radius $\frac{1}{2}$. The value $\frac{1}{2}$ is the best possible.

Gong and Liu [**GL**] gave some partial results about the growth theorem of the normalized biholomorphic convex mappings on the Reinhardt domain

$$B_p = \left\{ z = (z_1, \ldots, z_n) \;\middle|\; \|z\|_p = \left(\sum_{i=1}^{n} |z_i|^p \right)^{1/p} < 1 \right\}, \qquad p \geqslant 1.$$

Let $f(z) = (f_1(z), \ldots, f_n(z)) : B_p \to \mathbb{C}^n$ be a normalized biholomorphic convex mapping. Then

$$\|f(z)\|_p \leqslant \frac{\|z\|_p}{1 - \|z\|_p}, \qquad z \in B_p,$$

where $\|f(z)\|_p = \left(\sum_{i=1}^{n} |f_i(z)|^p \right)^{1/p}$. It implies that

$$|f(z)| \leqslant \frac{|z|}{1 - \|z\|_p}, \qquad z \in B_p,$$

These two inequalities are equivalent.

The paper cited above also claims some partial results about the growth theorem of the normalized biholomorphic convex mappings on the Reinhardt domain

$$D_p = \left\{ z = (z_1, \ldots, z_n) \in \mathbb{C}^n \mid |z_1|^{p_1} + \cdots + |z_n|^{p_n} < 1 \right\},$$

$p_1 \geqslant p_2 \geqslant \cdots \geqslant p_n \geqslant 1$.

Let $f(z) : D_p \to \mathbb{C}^n$ be a normalized biholomorphic convex mapping. Then

$$|f(z)| \leqslant \frac{\|z\|_p}{1 - \|z\|_p};$$

when $2 \geqslant p_1 \geqslant \cdots \geqslant p_n \geqslant 1$;

$$|f(z)| \leqslant n^{\frac{p_1-2}{2p_1}} \frac{\|z\|_p}{1 - \|z\|_p};$$

when $p_1 \geqslant p_2 \geqslant \cdots \geqslant p_n \geqslant 2$;

$$|f(z)| \leqslant \left(\ell^{\frac{p_1-2}{p_1}} + 1 \right)^{\frac{1}{2}} \frac{\|z\|_p}{1 - \|z\|_p};$$

when $p_1 \geqslant \cdots \geqslant p_\ell \geqslant 2 \geqslant p_{\ell+1} \geqslant \cdots \geqslant p_n \geqslant 1$, where $\|z\|_p = \left(\sum_{i=1}^{n} |z_i|^{p_i} \right)^{1/p_1}$.

The family of biholomorphic convex mappings is a family of $\mathcal{A}$-invariant mappings. One can apply all the results in section II.1 and II.3 to this family.

In section II.1, we discussed the distortion theorem of normalized biholomorphic convex mappings and concluded

$$\frac{(1 - |z|)^{C(\mathcal{S}) - \frac{n+1}{2}}}{(1 + |z|)^{C(\mathcal{S}) + \frac{n+1}{2}}} \leqslant |\det J_f(z)| \leqslant \frac{(1 + |z|)^{C(\mathcal{S}) - \frac{n+1}{2}}}{(1 - |z|)^{C(\mathcal{S}) + \frac{n+1}{2}}}$$

where $\frac{n+1}{2} \leqslant C(\mathcal{S}) \leqslant 1 + \frac{\sqrt{2}(n-1)}{2}$. There it is conjectured that $C(\mathcal{S}) = \frac{n+1}{2}$. This conjecture remains to be verified. But we have the following precise matrix form of distortion theorem of normalized biholomorphic convex mappings on the unit ball in $\mathbb{C}^n$. Gong, Wang and Yu proved the following result in [**GWY5**]:

If $f(z)$ is a normalized biholomorphic convex mapping on B^n which maps B^n to $\mathbb{C}^n$, then

$$\left(\frac{1-|z|}{1+|z|}\right)^2 G \leqslant J_f(z)\overline{J_f(z)}' \leqslant \left(\frac{1+|z|}{1-|z|}\right)^2 G,$$

where G is the Hermitian matrix $\left(\dfrac{(1-|z|^2)\delta_{ij}+\bar{z}^i z^j}{(1-|z|^2)^2}\right)_{1\leqslant i,j\leqslant n}$. The inequalities are precise.

The proof of theorem is based upon a Schwarz type lemma.

Moreover, we state two geometric properties of biholomorphic convex mappings.

Let $f(z)$ be the normalized biholomorphic convex mappings on B^n which maps B^n to $\mathbb{C}^n$, and $S_r = \{z \in \mathbb{C}^n \mid |z| = r < 1\}$. Then

(i)
$$\frac{(1-|z|)^5}{|z|(1+|z|)^3} \leqslant \text{ principal curvature of } f(S_r) \leqslant \frac{(1+|z|)^4}{|z|(1-|z|)^4};$$

(ii)
$$n\omega_{2n-1}\frac{1-|z|}{2|z|(1+|z|)}\int_0^{|z|}\frac{(1-x)^{n+1}x^{2n-1}dx}{(1+x)^{3n+3}} \leqslant \text{ volume of } f(S_r)$$
$$\leqslant n\omega_{2n-1}\frac{1+|z|}{2|z|(1-|z|)}\int_0^{|z|}\frac{(1+x)^{n+1}x^{2n-1}dx}{(1-x)^{3n+3}};$$

where ω_{2n-1} is the volume of the unit ball in $\mathbb{C}^n$.

3. Growth theorem for biholomorphic convex mappings on bounded symmetric domains.

Taishun Liu [LiuT2] proved the growth theorems and the covering $\frac{1}{2}$-theorems about the normalized biholomorphic convex mappings on bounded symmetric domains.

If $f(z) : \mathcal{R}_K \to \mathbb{C}^{\dim \mathcal{R}_K}$ is a normalized biholomorphic convex mapping on $\mathcal{R}_K$ $(K = I, II, III, IV)$, then, for any $Z \in \mathcal{R}_K$,

$$\frac{\|Z\|_K}{1+\|Z\|_K} \leqslant \|f(Z)\|_K \leqslant \frac{\|Z\|_K}{1-\|Z\|_K},$$

where $\|\cdot\|_K$ $(K = I, II, III, IV)$ is the norm of the element of $\mathcal{R}_K$ (cf. section I.2). The estimate is sharp. For example, the equalities hold for the following normalized biholomorphic convex mappings:

$$f(Z) = f_0(Z) = Z\left[I - \begin{pmatrix} z_{11} & \cdots & z_{1n} \\ 0 & \cdots & 0 \\ \vdots & \ddots & \vdots \\ 0 & \cdots & 0 \end{pmatrix}\right]^{-1}$$

in the case $\mathcal{R}_I$; $f(\bar{Z}) = f_0(Z)$, $Z = Z'$ in the case $\mathcal{R}_{II}$;

$$f(Z) = Z \left[I + \begin{pmatrix} 0 & 1 & 0 & \cdots & 0 \\ -1 & 0 & 0 & \cdots & 0 \\ \vdots & \vdots & \vdots & \ddots & \vdots \\ 0 & 0 & 0 & \cdots & 0 \end{pmatrix} Z \right]^{-1}, \qquad Z' = -Z$$

in the case of $\mathcal{R}_{III}$,

$$f(z) = (1 - 2z_1 + zz')^{-1}(z_1 - zz', z_2, \ldots, z_n)$$

in the case $\mathcal{R}_{IV}$.

As a consequence, we have the covering $\frac{1}{2}$-theorem.

If $f(Z) : \mathcal{R}_K \to \mathbb{C}^{\dim \mathcal{R}_K}$ is a normalized biholomorphic convex mapping on $\mathcal{R}_K$ ($K = I, II, III, IV$), then $f(\mathcal{R}_K)$ contains $\frac{1}{2}\mathcal{R}_K$.

According to a result in Gong, Wang and Yu [**GWY2**], the previous growth theorem is equivalent to another growth theorem, which states that if $f(Z) :$ $\mathcal{R}_K \to \mathbb{C}^{\dim \mathcal{R}_K}$ is a normalized biholomorphic convex mapping on $\mathcal{R}_K$ ($K = I, II, III, IV$), then for any $Z \in \mathcal{R}_K$,

$$\frac{|Z|}{1 + \|Z\|_K} \leqslant |f(Z)| \leqslant \frac{|Z|}{1 - \|Z\|_K},$$

where $|Z| = (\operatorname{tr} Z\bar{Z}')^{\frac{1}{2}}$, $|f(Z)| = (\operatorname{tr} f(Z)\overline{f(Z)}')^{\frac{1}{2}}$. The estimate is sharp as can be shown by using the above examples.

From the covering $\frac{1}{2}$-theorem, we have another kind of covering theorem.

Under the same assumption as above, $f(\mathcal{R}_I)$ contains $\frac{1}{2}B^{mn}$; $f(\mathcal{R}_{II})$ contains $\frac{1}{2}B^{n^2} \cap V_{II}$; $f(\mathcal{R}_{III})$ contains $\frac{\sqrt{2}}{2}B^{n^2} \cap V_{III}$; $f(\mathcal{R}_{IV})$ contains $\frac{\sqrt{2}}{4}B^n$; where V_{II}, V_{III} are defined at section III.4.

4. Distortion theorem for biholomorphic convex mappings on bounded non-symmetric transitive domains.

In the previous sections we have already discussed the biholomorphic convex mappings on bounded symmetric domains. Using the result in section II.2, Gong and Zheng [**GZ5**] obtained some estimate of the biholomorphic convex mappings on bounded non-symmetric transitive domains as follows.

Let $M \subset \mathbb{C}^n$ be a bounded nonsymmetric transitive domain with $0 \in M$ which is the Harish-Chandra canonical realization $\xi : G/K \to M$ of a homogeneous Kähler space G/K, where G is a Lie group consisting of some holomorphic automorphisms of M and acts transitively on M, K is the isotropy group of G which leaves 0 fixed. Let $f(z) : M \to \mathbb{C}^n$ be a normalized biholomorphic convex mapping on M. Then for any $z \in M$, the inequalities

$$\left(\frac{K(z, \bar{z})}{K(0, 0)} \right)^{\frac{1}{2}} e^{-2A\rho(z)} \leqslant |\det J_f(z)| \leqslant \left(\frac{K(z, \bar{z})}{K(0, 0)} \right)^{\frac{1}{2}} e^{2A\rho(z)}$$

holds, where K is the Bergman kernel function, $\rho(z)$ is the Bergman distance from 0 to z, and A is a constant satisfying the inequality

$$0 < A \leqslant B_M \left(n^{3/2} r_1^{-2} r_2 + \left(\sum_{\ell=1}^{n} |C_{\ell,M}|^2 \right)^{\frac{1}{2}} \right)$$

where B_M is a positive constant depending only on M, $n = \dim M$, r_1 is the radius of the largest ball centered at the origin in M, r_2 is the radius of the smallest ball centered at the origin containing M, and $C_{\ell,M} = K(0,0)^{-1} \frac{\partial}{\partial z_l} K(z,\bar{z}) \Big|_{z=0}$.

We also made the following conjecture. If M and f are as above, then for any $z \in M$, the inequalities

$$\left(\frac{K(z,\bar{z})}{K(0,0)} \right)^{\frac{1}{2}} \inf_{b \to B} \lim_{z_1 \to b} \left(\frac{K(z_1,\bar{z}_1)}{K(z_2,\bar{z}_2)} \right)^{\frac{1}{2}} \leqslant |\det J_f(z)|$$

$$\leqslant \left(\frac{K(z,\bar{z})}{K(0,0)} \right)^{\frac{1}{2}} \sup_{b \to B} \lim_{z_1 \to b} \left(\frac{K(z_1,\bar{z}_1)}{K(z_2,\bar{z}_2)} \right)^{\frac{1}{2}}$$

hold, where $z_1 = \xi(g \cdot O)$, $z_2 = \phi_g(z)$, $g \in G$, $O = eK$ is the coset of the unit element e in G/K, ϕ_g is the automorphism corresponding to g, z_1 approaches to b along a geodesic, and B is the Silov boundary of M.

5. Distortion theorem for locally biholomorphic convex mappings on bounded symmetric domains.

Note that the expression for the determinant of the Jacobian $\det J_f(z)$ of biholomorphic mapping on transitive domains at section II.2 is still valid if the mapping is a locally biholomorphic mapping. With this observation, Gong and Zheng [GZ5] obtained some results about the distortion theorems for locally biholomorphic convex mappings on bounded symmetric domains.

Let $M \subset \mathbb{C}^n$ be a bounded symmetric domain with $0 \in M$ and $f(z) : M \to \mathbb{C}^n$ be a locally biholomorphic mapping. The mapping $f(z)$ is called Ω-uniform locally biholomorphic convex mapping if there exists a open set $\Omega \subset M$ with $0 \in \Omega$, such that for any $z \in M$, $\phi_z \in G$, $f \circ \phi_z$ is biholomorphic convex mapping on Ω, where G is the group of holomorphic automorphisms of M.

Let $f(z) : M \to \mathbb{C}^n$ be a Ω-uniform locally biholomorphic convex mapping on M. If Ω contains the largest ball centered at the origin with a radius r, and is in the smallest ball centered at the origin with a radius R, then

$$\left(\frac{K(z,\bar{z})}{K(0,0)} \right)^{\frac{1}{2}} \prod_{j=1}^{q} \left(\frac{1 - |\tanh x_j|}{1 + |\tanh x_j|} \right)^{A_j} \leqslant |\det J_f(z)|$$

$$\leqslant \left(\frac{K(z,\bar{z})}{K(0,0)} \right)^{\frac{1}{2}} \prod_{j=1}^{q} \left(\frac{1 + |\tanh x_j|}{1 - |\tanh x_j|} \right)^{A_j},$$

where x_j $(j = 1, \ldots, q)$ are as defined in section I.3, and A_j $(j = 1, \ldots, q)$ are
constants which satisfy the inequalities

$$\frac{1}{4} \sum_{p=q+1}^{2n} |\alpha_p(X_j)| + \frac{1}{2} \leqslant A_j \leqslant nr^{-2}R$$

where $\alpha_{q+1}, \ldots, \alpha_{2n}$ are the positive roots (counting the multiplicity) of the
adjoint representations of $\mathfrak{a}$ at $\mathfrak{g}$ (cf. section II.3 for the definitions of $\mathfrak{a}$ and $\mathfrak{g}$).

V. Schwarzian derivatives

1. Schwarzian derivatives on classical domains.

The importance of the Schwarzian derivative of one complex variable is well
known. Recall that if $f(z)$ is an analytic function in the unit disc $|z| < 1$ with
$f' \neq 0$, then the Schwarzian derivative of $f(z)$ is defined as

$$S(f) = \frac{f'''}{f'} - \frac{3}{2}\left(\frac{f''}{f'}\right)^2 = \left(\frac{f''}{f'}\right)' - \frac{1}{2}\left(\frac{f''}{f'}\right)^2.$$

There are two fundamental properties of Schwarzian derivative of one complex
variable.

1) The Schwarzian derivative of an analytic function $w = f(z)$ in $|z| < 1$ is
invariant if w is composed with a fractional linear transformation.
2) If $f(z)$ in $|z| < 1$ is annihilated by Schwarzian derivative, then $f(z)$ must
be a linear fractional transformation.

How can the idea of Schwarzian derivative be extended to the case of several
complex variables? In 1988, FitzGerald and Gong [FG1] extended the concept
to the classical domain of type $\mathcal{R}_I : I - Z\bar{Z}' > 0$, $Z = Z^{(n)}$, and obtained some
corresponding properties.

Let $W = W(Z) = (w_{jk})_{1 \leqslant j,k \leqslant n} : \mathcal{R}_I \to \mathbb{C}^{n^2}$ be a holomorphic mapping on
$\mathcal{R}_I$, $\Lambda = (\lambda_{jk})_{1 \leqslant j,k \leqslant n}$ be a nonsingular matrix,

$$\frac{\partial}{\partial z} = \left(\frac{\partial}{\partial z_{11}}, \ldots, \frac{\partial}{\partial z_{1n}}, \ldots, \frac{\partial}{\partial z_{n1}}, \ldots, \frac{\partial}{\partial z_{nn}}\right),$$

$\lambda = (\lambda_{11}, \ldots, \lambda_{1n}, \ldots, \lambda_{n1}, \ldots, \lambda_{nn})$, and $\left(\frac{\partial}{\partial z}\right)^{[k]}$, $\lambda^{[k]}$ denote the k-th Kro-
necker products of $\frac{\partial}{\partial z}$ and λ respectively. The directional derivative of W
along Λ is $D_\Lambda W = \lambda\left(\frac{\partial}{\partial z}\right)' W$. The directional derivatives of high order are
$D_\Lambda^k W = \lambda^{[k]}\left(\frac{\partial}{\partial z}\right)^{[k]} W$, $k = 2, 3, \ldots$. If $D_\Lambda W$ is nonsingular, then one can
define the Schwarzian derivative of W along the direction Λ as

$$\{W, Z\}_\Lambda = D_\Lambda^3 W(D_\Lambda W)^{-1} - \frac{3}{2}(D_\Lambda^2 W)(D_\Lambda W)^{-1}(D_\Lambda^2 W)(D_\Lambda W)^{-1}.$$

We proved the following two important properties of the Schwarzian derivative.

(1) $\{W, Z\}_\Lambda$ is invariant up to similarity when W is composed with any element of the group of holomorphic automorphisms of $\mathcal{R}_I$.

(2) $\{W, Z\}_\Lambda = 0$ for any $Z \in \mathcal{R}_I$ and any nonsingular matrix Λ if and only if W is

$$W(Z) = (w_{ij}(Z))_{1 \leqslant i, j \leqslant n}$$

with

$$w_{ij}(Z) = c_{ij} + \operatorname{tr}\left(C_{ij} Z(I - L(Z))^{-1}\right),$$

or

$$w_{ij}(Z) = c_{ij} + \operatorname{tr}\left(C_{ij}(I - L(Z))^{-1} Z\right),$$

$i, j = 1, 2, \ldots, n$. Here c_{ij} are constants, and C_{ij} are $n \times n$ constant matrices, and $L(Z)$ is a $n \times n$ matrix whose entries are homogeneous linear function of z_{ij}, $i, j = 1, 2, \ldots, n$.

The following algebraic theorem is needed in proving (2).

Let $Z = (z_{ij})$ be an $n \times n$ matrix of complex variables. Let W be the adjoint of Z. Consider a row vector A of n entries, with each entry being a homogeneous polynomial of degree two in the variables z_{ij}. And consider a column vector B of n entries, with each entry being a homogeneous polynomial of degree two in the variables z_{ij}. If AWB is divisible by the polynomial $\det Z$, then either each entry of the vector AW is divisible by $\det Z$, or each entry of the vector WB is divisible by $\det Z$. Consequently, either there exists a row vector L with first order entries in the z_{ij} variables such that $A = LZ$ or a column vector L with first order entries such that $B = ZL$.

In the high dimensional case, there exists another invariant up to similarity.

Let $W = W(Z)$ be a holomorphic mapping which maps $\mathcal{R}_I$ into $\mathcal{R}_I$. Then $[W, Z]_{\Lambda, \Pi} = D_\Lambda W (D_\Pi W)^{-1}$ is invariant up to similarity when W is composed with any element of the group of holomorphic automorphisms of $\mathcal{R}_I$ where Λ and Π are any two $n \times n$ matrices when $D_\Lambda W$ and the inverse of $D_\Pi W$ exist. When $n = 1$, this invariant does not exist.

Similarly, we can define the Schwarzian derivative and $[W, Z]_{\Lambda, \Pi}$ of holomorphic mappings on $\mathcal{R}_{II}$, $\mathcal{R}_{III}$ and the corresponding real classical domains, and obtain the corresponding properties.

2. Schwarzian curvatures of holomorphic curves.

The Schwarzian derivative of one complex variables, however, has a more direct geometric interpretation. Flanders [**F**] interpreted the Schwarzian of a C^3 function $w = f(z)$ as a natural invariant of an equivalence problem for curves in the projective line. Weiqi Gao [**G**] extends it to the high dimensions.

Let $\phi : D \to \mathbb{P}^n$ be a curve in the n-dimensional projective space (over $\mathbb{C}$), where D is the unit disk in $\mathbb{C}$. We may lift it to a curve in $\mathbb{C}^{n+1}$, $f : D \to \mathbb{C}^{n+1}$. Two such liftings f_1 and f_2 are equivalent if and only if $f_1 = \lambda f_2$ for some nonzero function λ.

The moving frame on the curve f is constructed as follows. The normal vector is $\nu = \lambda f$, where $\lambda \neq 0$. The tangent vectors are derivatives of ν, $e_1 = \nu'$, $e_2 = e_1'$, $\cdots$, $e_n = e_{n-1}'$. We can choose λ so that $\lambda^{n+1} [f, f', \ldots, f^{(n)}] = 1$. Here the brackets indicate the determinant. We have $e_n' = \kappa_0 \nu + \kappa_1 e_1 + \cdots + \kappa_{n-1} e_{n-1}$. We call $\kappa_0, \kappa_1, \ldots, \kappa_{n-1}$ the Schwarzian curvatures of the curve ϕ.

The quantities $\kappa_0, \kappa_1, \ldots, \kappa_{n-1}$ are invariant under projective transformation in $\mathbb{P}^n$, or equivalently, under affine transformations of $\mathbb{C}^{n+1}$.

All the Schwarzian curvatures of a curve ϕ vanish, if and only if ϕ is a n-th degree polynomial fractional map into $\mathbb{P}^n$, i.e.,

$$\phi = \left(\frac{P_1(z)}{P_0(z)}, \ldots, \frac{P_n(z)}{P_0(z)} \right) .$$

More general, all the Schwarzian curvatures of a curve ϕ are constants, if and only if

$$\phi = \left(\frac{L_1}{L_0}, \ldots, \frac{L_n}{L_0} \right) ,$$

where L's are linear combinations of the $n+1$ linearly independent solutions of the $(n+1)$-st order differential equation with constant coefficients

$$y^{(n+1)} = \kappa_0 y + \kappa_1 y' + \cdots + \kappa_{n-1} y^{(n-1)}.$$

Let $k_0(z), k_1(z), \ldots, k_{n-1}(z)$ be analytic functions on D. Then there are curves with $\kappa_0 = k_0(z), \kappa_1 = k_1(z), \ldots, \kappa_{n-1} = k_{n-1}(z)$.

We can express all the curvatures κ_j, $j = 0, 1, \ldots, n-1$, as functions of $\lambda, \lambda', \ldots, \lambda^{(n+1)}$ and $g_0, g_1, \ldots, g_n$ where $f^{(n+1)} = g_0 f + g_1 f' + \cdots + g_n f^{(n)}$. When $n = 2$, $\phi = x(z)$ and $f = (1, x)$, we have

$$\kappa_0 = \frac{\lambda'}{\lambda} = \frac{3}{4} \left(\frac{x''}{x'} \right)^2 - \frac{1}{2} \left(\frac{x'''}{x'} \right) ,$$

a constant multiple of the traditional Schwarzian derivative of $x(z)$.

When $n = 3$, $\phi = (x(z), y(z))$ and $f = (1, x, y)$, we have

$$\kappa_1 = \frac{3\lambda''}{\lambda} + g_1, \qquad \kappa_0 = -\frac{\lambda'}{\lambda} \kappa_1 + \frac{\lambda'''}{\lambda} .$$

Let $z = z(w)$ be a change of coordinates in D. Then the Schwarzian curvatures are not invariant. Instead, they obey a set of transformation rules.

Denote the derivatives in z by primes and the derivatives in w by dots. And denote the quantities associated with the w variables by a tilde. Then the transformation formulas is given by

$$\tilde{\kappa}_0 = \dot{z}^2 \kappa_0 + Sz$$

when $n = 1$, and

$$\tilde{\kappa}_1 = \dot{z}^2 \kappa_1 + 4Sz, \qquad \tilde{\kappa}_0 = \dot{z}^3 \kappa_0 + \dot{z}\ddot{z}\kappa_1 + 2(Sz)'$$

when $n = 2$, where

$$Sz = \frac{3}{4}\left(\frac{\ddot{z}}{\dot{z}}\right)^2 - \frac{1}{2}\left(\frac{\dddot{z}}{\dot{z}}\right)$$

is a constant multiple of the traditional Schwarzian derivative of z.

We obtain the transformation formulas of the κ's in higher dimensions, when we use the chain rule polynomials. In particular,

$$\tilde{\kappa}_{n-1} = \dot{z}^2 \kappa_{n-1} + \binom{n+2}{3} Sz.$$

Recently, there are some other important extensions of Schwarzian derivatives. K. Carne [CarK] and B. Osgood and D. Stowe [OS1, OS2] extended the Schwarzian derivatives in the point view of the conformal mappings between Riemannian manifolds.

VI. The Bloch constant of holomorphic mappings on bounded symmetric domains

1. The Bloch constant of holomorphic mappings on the unit ball.

Assume that F is a holomorphic function in the unit disc D of the complex plane. The function F is called a Bloch function if and only if its Bloch norm given by $\|F\| = \sup\{(1 - |z|^2)|F'(z)| \mid |z| < 1\}$ is finite. For a point a in D, let $r(a, F)$ denote the radius of the largest schlicht disk on the Riemann surface $F(D)$ centered at $F(a)$. (A schlicht disk on $F(D)$ centered at $F(a)$ means that F maps an open subset of D containing a conformally onto this disc). Let $r(F)$ be the supremum of $r(a, F)$ for all a in D. The Bloch constant B is defined to be the infimum of all $r(F)$, where F is normalized by $F'(0) = 1$.

In 1924, Bloch proved that B is positive. The exact value of the Bloch constant is still unknown. In 1926, E. Landau showed that $0.396 < B < 0.555$. In 1937, L. V. Ahlfors and H. Grunsky improved the upper bound of the Bloch constant as $B \leqslant B_0 = \dfrac{\Gamma\left(\frac{1}{3}\right)\Gamma\left(\frac{11}{12}\right)}{\left(1 + 3^{\frac{1}{2}}\right)^{\frac{1}{2}}\Gamma\left(\frac{1}{4}\right)} < 0.4719$, and conjectured B_0 to be the exact value of B. In 1938, L. V. Ahlfors [A] published a very important paper, in which he established the lower bound $B \geqslant \frac{\sqrt{3}}{4}$. In 1962, M. Heins [Hei] eliminated the equality sign of the estimation. This estimation for the lower bound of the Bloch constant is accepted for half century until 1988, when M. Bonk [Bon] made a numerically small improvement of the lower bound on the Bloch constant to $B \geqslant \frac{\sqrt{3}}{4} + 10^{-14}$ by making use a distortion theorem obtained by himself which can be stated as follows. If $\|F\| = 1$ and $F'(0) = 1$, then $\operatorname{Re} F'(z) \geqslant (1 - \sqrt{3}|z|)\left(1 - \dfrac{|z|}{\sqrt{3}}\right)^{-3}$, when $|z| \leqslant \frac{1}{\sqrt{3}}$. Recently D. Minda [M] gave an elementary and geometric proof of Bonk's result.

The study of the Bloch constant of mappings in high dimensions was started by S. Bochner [Boc]. In 1948, he proved the following theorem.

Assume that $f(x) = (f_1(x), \ldots, f_n(x))$ $(x = (x_1, \ldots, x_n) \in \mathbb{R}^n)$ satisfies

(1) the Laplace equations

$$\frac{\partial^2 f_i}{\partial x_1^2} + \cdots + \frac{\partial^2 f_i}{\partial x_n^2} = 0, \qquad i = 1, 2, \ldots, n$$

in the unit ball $x_1^2 + \cdots + x_n^2 \leqslant 1$,

(2) $\det J_f(0) = 1$ and

(3) $\operatorname{tr}(J_f(x)J_f(x)') \leqslant K|\det J_f(x)|^{n/2}$ where K is a positive constant.

Then there exists an open set S in the unit ball, such that $f(x)$ maps S one to one to a ball with radius $R_0 = R_0(n, K)$ in $\mathbb{R}^n$.

In 1951, Takahashi extended Bochner's result to the holomorphic mappings on the unit ball in $\mathbb{C}^n$. In 1967, H. Wu [**W**] pointed out that the Bloch constant of holomorphic mappings on domains in $\mathbb{C}^n$ does not exist in the absence of any restrictions on the mappings. He also defined the quasicomformal mappings in several complex variables, and got several interesting results. Since then there have been many results about the Bloch constant of holomorphic mappings with various restrictions on the Jacobian.

In 1990, Xiangyang Liu [**LiuX**] extended the Bonk's distortion theorem to the unit ball in $\mathbb{C}^n$. He discussed the family $\mathcal{B}(K)$ of holomorphic mappings on the unit ball B^n in $\mathbb{C}^n$. A holomorphic mapping f on B^n is said to belong to $\mathcal{B}(K)$ if $\|f\| \leqslant K$, $1 \leqslant K < \infty$, where $\|f\| = \sup\{|(f \circ \phi)'(0)| \mid \phi \in$ group of holomorphic automorphisms of $B^n\}$. Liu proved the following result.

If $f \in \mathcal{B}(K)$ with $\det J_f(0) = 1$, then the Bloch constant of $\mathcal{B}(K)$ satisfies the inequalities

$$K^{1-n} \geqslant B(\mathcal{B}(K)) \geqslant C(K, n)$$

where

$$C(K, n) = K^{1-n} \int_0^{(n+1)^{-\frac{1}{2}}} (1 - t^2)^{n-1}(1 - (n+2)^{\frac{1}{2}} t)$$

$$(1 - (n+2)^{-\frac{1}{2}} t)^{-(n+2)} dt$$

$$\geqslant K^{1-n} \frac{\sqrt{n+2}}{en} \left(\left(1 + \frac{1}{n+1}\right)^{n+1} - 2 \right).$$

This result coincides with Ahlfors' result $B \geqslant \frac{\sqrt{3}}{4}$ when $n = 1$.

2. The Bloch constant of holomorphic mappings on classical domains.

In 1990, FitzGerald and Gong [**FG2**] extended Liu's result [**LiuX**] in the previous section to classical domains.

Let M be the bounded transitive domain and $\operatorname{Aut}(M)$ be the group of holomorphic automorphisms of M. A holomorphic mapping $f(z)$ is called a Bloch mapping if $F_f = \{g \mid g(z) = f(\phi(z)) - f(\phi(0))$, all $\phi \in \operatorname{Aut}(M)\}$ forms a normal family, where $z \in M$, $f(z) : M \to \mathbb{C}^n$. The norm of f is defined as

$\|f\|_M = \sup\{\|J_{f \circ \phi}(0)\| \mid \phi \in \text{Aut}(M)\}$. A holomorphic mapping $f : M \to \mathbb{C}^n$ with $f(0) = 0$ is a Bloch mapping if and only if $\|f\|_M < \infty$. Let $D(f)$ denote $\sup_{z \in M} \dfrac{|\det J_f(z)|}{\sqrt{K(z, \bar{z})/K(0,0)}}$. Then $D(f)$ is invariant under $\text{Aut}(M)$. We extended Bonk's theorem to the classical domains. Here we just state the result for the classical domain $\mathcal{R}_I : I - Z\bar{Z}' > 0$, $Z = Z^{(m,n)}$.

Let f be the Bloch mapping with $D(f) = 1$ and $\det J_f(0) = 1$. Then

$$\text{Re}\left\{\det J_f(Z)\right\} \geqslant \left(1 - \sqrt{m^2 + mn + 1}\,\|Z\|\right)\left(1 - \frac{\|Z\|}{\sqrt{m^2 + mn + 1}}\right)^{-(m^2 + mn + 1)}$$

for $\|Z\| \leqslant \dfrac{2\sqrt{m^2 + mn + 1}}{m^2 + mn + 2}$, where $\|\cdot\|$ is the norm of the matrix.

By making use the above result, we obtained the estimation of the Bloch constant of family $\mathcal{B}(K)$. A holomorphic mapping $f(Z) : \mathcal{R}_I \to \mathbb{C}^{mn}$ is said to belong to $\mathcal{B}(K)$ if $\|f\|_{\mathcal{R}_I} \leqslant K$ and $\det J_f(0) = 1$.

The Bloch constant B of $\mathcal{B}(K)$ satisfies the following inequality.

$$B \geqslant K^{1-m} \int_0^{(m+mn+1)^{-\frac{1}{2}}} \left(1 - (m^2 + mn + 1)^{\frac{1}{2}}t\right)$$
$$(1 - t^2)^{mn-1}\left(1 - \frac{t}{\sqrt{m^2 + mn + 1}}\right)^{-(m^2 + mn + 1)} dt$$

and $K^{1-mn} \geqslant B$. It coincides with Xiangyang Liu's result when $m = 1$.

3. The Bloch constant of holomorphic mappings on bounded symmetric domain.

Gong and Yan [**GY**] extended the previous results to the bounded symmetric domains as follows.

Let D be an irreducible bounded symmetric domain containing the origin with standard Harish-Chandra realization. Let G be the connected component of $\text{Aut}(D)$, and $\mathfrak{g}$ be the Lie algebra of G. Let K be an isotropy group of G which leaves the origin fixed and $\mathfrak{k}$ be the maximal compact subalgebra of $\mathfrak{g}$ which corresponds to K. Then $\mathfrak{g}$ is a real simple Lie algebra, and has the Cartan decomposition $\mathfrak{g} = \mathfrak{k} + \mathfrak{p}$. Let $\mathfrak{h}$ be Cartan subalgebra of $\mathfrak{k}$. The $\mathfrak{h}^{\mathbb{C}}$-root of $\mathfrak{g}^{\mathbb{C}}$ that are also root of $\mathfrak{k}^{\mathbb{C}}$ are called compact roots. We denote by Φ^+ the set of positive non-compact roots. Denoting by τ the conjugation with respect to the real form $\mathfrak{k} + i\mathfrak{p}$, we consider a basis of root vectors e_α such that $\tau e_\alpha = -e_{-\alpha}$, $[e_\alpha, e_{-\alpha}] = h_\alpha$, $[h_\alpha, e_{\pm\alpha}] = \pm 2e_{\pm\alpha}$. Setting $\mathfrak{p}^{\pm} = \sum_{\alpha \in \Phi^+} \mathbb{C}e_{\pm\alpha}$, we have $\mathfrak{g}^{\mathbb{C}} = \mathfrak{p}^- + \mathfrak{k}^{\mathbb{C}} + \mathfrak{p}^+$ as a vector space direct sum. We denote by $\gamma_1, \ldots, \gamma_r$ the strong orthogonal roots of Harish-Chandra, where r is the real rank of $\mathfrak{g}$. We simply write $e_j = e_{\gamma_j}$, $h_j = h_{\gamma_j}$, $1 \leqslant j \leqslant r$ and $e = \sum_{j=1}^r e_j$. It is known that in Harish-Chandra realization, D is a bounded symmetric domain in $\mathfrak{p}^+$, and the isotropy subgroup K of G at the origin acts on D by unitary transformations.

The restricted roots are given by $+\frac{1}{2}(\gamma_j + \gamma_k)$, $+\gamma_j$, $+\frac{1}{2}\gamma_j$ ($1 \leqslant j, k \leqslant r$) with respective multiplicities a, 1 and $2b$ (independent of j, k). Put $p = (r-1)a+b+2$.

By using the results of Faraut and Koranyi [**FK**], we have been able to extend the Bonk theorem to bounded symmetric domains.

Let D be an irreducible bounded symmetric domain with standard Harish-Chandra realization and $F(z)$ be a Bloch mapping on D with $D(F) = 1$, $\det \dfrac{\partial F}{\partial z}(0) = 1$, $z = k \sum_{i=1}^{r} \mu_i e_i \in D$, $k \in K$. Then

$$\mathrm{Re}\left\{\det \frac{\partial F}{\partial z}(z)\right\} \geqslant \frac{1 - \sqrt{pr+1}\,\|z\|}{\left(1 - \dfrac{\|z\|}{\sqrt{pr+1}}\right)^{pr+1}}$$

when $\|z\| < \frac{2\sqrt{pr+1}}{pr+2}$, and $\|z\| = \mu_1$.

Using this extended Bonk theorem, we can get the following estimation of the Bloch constant.

Let $F(z)$ be a Bloch mapping on D, $\|F\|_B \leqslant K$, $\det J_f(0) = 1$. Then the Bloch constant B_K satisfies

$$B_K \geqslant K^{1-n} \int_0^{\frac{1}{\sqrt{pr+1}}} \frac{(1 - \sqrt{pr+1}\,t)(1 - t^2)^{n-1}}{\left(1 - \dfrac{t}{\sqrt{pr+1}}\right)^{pr+1}}\,dt.$$

When $D = \mathcal{R}_I$, one has $p = m+n$, $r = m$, hence $(pr+1)^{\frac{1}{2}} = (m^2 + mn + 1)^{\frac{1}{2}}$. This coincide with the result in previous section. When $D = \mathcal{R}_{II}$, one has $p = n+1$, $r = n$, hence $(pr+1)^{\frac{1}{2}} = (n^2 + n + 1)^{\frac{1}{2}}$. When $D = \mathcal{R}_{III}$, one has $p = 2n-2$, $r = [\frac{1}{2}n]$, hence $(pr+1)^{\frac{1}{2}} = (n^2 - n + 1)^{\frac{1}{2}}$ when n is even; and $(pr+1)^{\frac{1}{2}} = (n^2 - 2n + 2)^{\frac{1}{2}}$ when n is odd. When $D = \mathcal{R}_{IV}$, one has $p = n$, $r = 2$, hence $(pr+1)^{\frac{1}{2}} = (2n+1)^{\frac{1}{2}}$. When D is the exceptional domain of dimension 16, one has $p = 12$, $r = 2$, hence $(pr+1)^{\frac{1}{2}} = 5$. When D is the exceptional domain of dimension 27, one has $p = 18$, $r = 3$, hence $(pr+1)^{\frac{1}{2}} = 7$. Substitute all these values in the previous formula, then we get the estimations of the Bloch constants on the corresponding bounded symmetrical domains.

References

[A]   Ahlfors, L. V., *An extension of Schwarz lemma*, Trans. Amer. Math. Soc. **43** (1938), 359–364.

[BFG1]   Barnard, R. W., FitzGerald, C. H. and Gong, Sheng, *Distortion theorem for biholomorphic mappings in* $\mathbb{C}^2$, Chin. Sci. Bull. **35** (1990), 353–356; Tran. Amer. Math. Soc. (to appear).

[BFG2]   ———, *The growth and $\frac{1}{4}$-theorems for starlike mappings in* $\mathbb{C}^n$, Chin. Sci. Bull. **34** (1989), 161–162; Pacific Jour. of Math. **150** (1991), 13–22.

[Boc]   Bochner, S., *Bloch theorem for real variables*, Bull. Amer. Math. Soc. **52** (1946), 715–719.

[Bon]   Bonk, M., *On Bloch's constant*, Proc. Amer. Math. Soc. **110** (1990), 889–894.

[CarK]   Carne, K., *The Schwarzian derivative for conformal maps*, J. Reine. Angew. Math. **408** (1990), 10–33.

[CarH] Cartan, H., *Sur la possibili té d'étendre aux fonctions de plusieurs variables complexes la theorie des fonctions univalents*, Lecons sur les Fonctions Univalents on Multivalents, by P. Montel, Gauthier-Villar, 1933, pp. 129–155.

[CR] Chen, Hong-bin and Ren, Fu-yao, *Univalence of holomorphic mappings and growth theorems for close-to-starlike mappings in finitely dimensional Banach space*, preprint (1991).

[DZ] Dong, Dao-zhan and Zhang, Wen-jun, *The growth and $\frac{1}{4}$-theorems for starlike mappings in a Banach space*, preprint (1990).

[FK] Faraut, J. and Koranyi, A., *Function spaces and reproducing kernels on bounded symmetric domains*, Journal of Functional Analysis **88** (1990), 64–89.

[FG1] FitzGerald, C. H. and Gong, Sheng, *Schwarz derivatives in several complex variables* I, Chin. Sci. Bull. **34** (1989), 391–392.

[FG2] _____, *The Bloch theorem in several complex variables*, Chin. Sci. Bull. **37** (1992), 406–408; Journal of Geometric Analysis (to appear).

[F] Flanders, H., *The Schwarzian derivative as a curvature*, J. Diff. Geom. **4** (1970), 515–519.

[G] Gao, Weiqi, *Schwarzian curvatures of holomorphic curves*, preprint (1992).

[GL] Gong, Sheng and Liu, Taishun, *The growth theorem of biholomorphic convex mappings on B_p*, Chin. Quar. Jour. of Math. **6** (1991), 78–82.

[GWY1] Gong, Sheng, Wang, Shikun and Yu, Qihuang, *The growth and $\frac{1}{4}$-theorems for starlike mappings in B_p*, Chin. Ann. of Math. **11** (1990), 100–104.

[GWY2] _____, *The growth theorems in several complex variables*, Chin. Science Bull. **36** (1991), 1148–1150.

[GWY3] _____, *A necessary and sufficient condition that biholomorphic mappings are starlike on Reinhardt domains*, Chin. Science Bull. **36** (1991), 1151–1153; Chin. Ann. of Math. **13B(1)** (1992), 95–104.

[GWY4] _____, *On biholomorphic starlike mappings in several complex variables*, Chin. Science Bull. **37** (1992).

[GWY5] _____, *Biholomorphic convex mappings of complex hyperball*, Pacific J. of Math. (to appear).

[GY] Gong, Sheng and Yan, Zhimin, *Bloch constant of holomorphic mappings on bounded symmetric domains*, Scientica Sinica (to appear).

[GZ1] Gong, Sheng and Zheng, Xue-an, *Distortion theorem for biholomorphic mappings in transitive domains* I, Chi. Sci. Bull. **35** (1990), 1167–1171; International Symposium in Memory of L. K. Hua, Vol. II, Springer-Verlag, 1991, pp. 111–122.

[GZ2] _____, *Distortion theorem for biholomorphic mappings in transitive domains* II, Chi. Sci. Bull. **35** (1990), 330–332; Chin. Ann. of Math. **13B** (1992), 91–104.

[GZ3] _____, *Distortion theorem for biholomorphic mappings in transitive domains* III, Chi. Sci. Bull. **35** (1990), 1214–1217; Chin. Ann. of Math. (to appear).

[GZ4] _____, *Distortion theorem for biholomorphic mappings in transitive domains* IV, preprint (1991).

[GZ5] _____, *Distortion theorem for biholomorphic mappings in transitive domains* V, preprint (1991).

[Hei] Heins, M., *On a class of conformal metrics*, Nagoya Math. J. **21** (1962), 1–60.

[Har] Harris, L. A., *Bounded symmetric homogeneous domains in infinite dimensional space*, Lecture Notes in Mathematics No. 364, 1973, pp. 13–40.

[Hua] Hua, Luo-Keng, *Harmonic Analysis of Functions of Several Complex Variables in the Classical Domains*, Transl. Amer. Math. Soc., vol. 6, 1963.

[Kik] Kikuchi, K., *Starlike and convex mappings in several complex variables*, Pacific Jour. of Math. **44** (1973), 569–580.

[Kim] Kim, D., *Complete domains with respect to the Caratheodory distance* II, Proc. Math. Soc. **53** (1975), 141–142.

[Kob] Kobayashi, S., *Hyperbolic Manifolds and Holomorphic Mappings*, Dekker, New York, 1970.

[LiuT1] Liu, Taishun, *The distortion theorem for biholomorphic mappings in $\mathbb{C}^n$*, Chin. Ann. of Math. (to appear).

[LiuT2]   ———, *The growth theorems, covering theorems and distortion theorems for biholo-morphic mappings on classical domains*, University of Science and Technology of China Thesis (1989).

[LiuX]   Liu, Xiangyang, *Bloch functions of several complex variables*, Pacific Journal of Math. **152** (1992), 347–363.

[M]   Minda, D., *The Bloch and Marden constants*, Proceeding of CMFT '89, Lecture Notes in Mathematics, No. 1435, 1990, pp. 131–142.

[MT]   Mok, Ngaiming and Tsai, I-Hsun, *Rigidity of convex realizations of irreducible bounded symmetric domains of rank ≥ 2*, preprint (1990).

[OS1]   Osgood, B. and Stowe, D., *The Schwarzian derivative and conformal mappings of Riemannian manifolds*, preprint (1989).

[OS2]   ———, *A generalization of Nehari's univalence criterion*, Comment. Math. Helv. **65** (1990), 234–242.

[Pfa1]   Pfaltzgraff, J. A., *Subordination chains and univalence of holomorphic mappings on $\mathbb{C}^n$*, Math. Ann. **210** (1974), 55–68.

[Pfa2]   ———, *Loewner theory in $\mathbb{C}^n$*, Abstracts of Papers Presented to the American Mathematical Society **11(66)** (January 1990), p. 46.

[PS]   Poletskii, E. A. and Shabat, B. V., *Invariant metrices, Several complex variables III* (Khenkin G. M., ed.), Encyclopaedia of mathematical Sciences, Vol. 9, Springer-Verlag, 1989, pp. 63–112.

[Pom1]   Pommerenke, Ch., *Linear-invariante Familien analytischer Funktionen I*, Math. Ann. **155** (1964), 108–154.

[Pom2]   ———, *Linear-invariante Familien analytischer Funktionen II*, Math. Ann. **156** (1964), 226–262.

[R]   Rudin, W., *Function Theory in the Unit Ball of $\mathbb{C}^n$*, Springer-Verlag, 1980.

[S1]   Suffridge, T. J., *The principle of subordination applied to function of several variables*, Pacific Jour. of Math. **33** (1970), 241–248.

[S2]   ———, *Starlike and convex maps in Banach spaces*, Pacific Jour. of Math. **46** (1973), 575–589.

[S3]   ———, *Starlikeness, convexity and other geometric properties of holomorphic maps in higher dimension*, Lecture Notes in Mathematics No. 599, 1976, pp. 146–159.

[S4]   ———, *Biholomorphic mappings of the ball onto convex domains*, Abstracts of Papers Presented to American Mathematical Society **11(66)** (January 1990), p. 46.

[T]   Thomas, C., *Convex mappings from the unit ball in $\mathbb{C}^n$ into $\mathbb{C}^n$*, Abstracts of Papers Presented to American Mathematical Society **11(66)** (January 1990), p. 46.

[W]   Wu, H., *Normal families of holomorphic mappings*, Acta Math. **119** (1967), 193–233.

DEPARTMENT OF MATHEMATICS, UNIVERSITY OF SCIENCE AND TECHNOLOGY OF CHINA, HEFEI, ANHUI, 230026, THE PEOPLE'S REPUBLIC OF CHINA

Contemporary Mathematics
Volume **142**, 1993

Global Lojasiewicz Inequality, Defect Relation and Applications of Holomorphic Curve Theory

SHANYU JI AND MIN RU

In this paper, some most recent research work related to algebraic and complex geometry contributed by the authors are surveyed.

1. Global Lojasiewicz Inequality

We recall the classical Lojasiewicz inequality (cf.[**M, theorem 4.1**]) for real-analytic sets: we let $d(x, S)$ denote the Euclidean distance from a point $x \in \mathbb{R}^n$ to a set $S \subset \mathbb{R}^n$. Suppose f is a real-analytic function on an open set Ω in $\mathbb{R}^n$, and let $Z(f) = \{x \in \Omega \mid f(x) = 0\}$. Then for any compact set $K \subset \Omega$, there exist constants C, $\alpha > 0$ such that

$$d(x, Z(f))^\alpha \leqslant C|f(x)|, \qquad \text{for all } x \in K.$$

In general α can be large. For complex polynomials, Brownawell [**B**] proved a global inequality: let $f_1, \ldots, f_k \in \mathbb{C}[z_1, \ldots, z_n]$ and let $D = \max_i d_i$, where $d_i = \deg f_i$ and $d_1 \leqslant \cdots \leqslant d_k$. Let $Z = \{z \in \mathbb{C}^n \mid f_1(z) = \cdots = f_k(z) = 0\}$. Then there is a constant $C > 0$ such that

$$\left(\frac{\min(\text{dist}(z, Z), 1)}{1 + \|z\|^2} \right)^{(n+1)^2 D^{\min(n,k)}} \leqslant C \cdot \max_i |f_i(z)|$$

for any $z \in \mathbb{C}^n$, where $\|z\|^2 = |z_1|^2 + \cdots + |z_n|^2$. The improvement of Brownawell's result was given by Ji, Kollàr and Shiffman:

1991 *Mathematics Subject Classification.* 32A22, 32H30, 32-02.

Partial financial support for the first author was provided by a University of Houston Initiation Research Grant and by the NSF under grant number DMS-8922760.

This paper is in final form and no version of it will be submitted for publication elsewhere.

THEOREM [**JKS**]. *Using the notation above, we have*

$$\left(\frac{\min(\mathrm{dist}(z,Z),1)}{1+\|z\|}\right)^{\bar{B}(n,d_1,\ldots,d_k)} \leqslant C \cdot \max_i |f_i(z)|$$

and

$$\left(\frac{\mathrm{dist}(z,Z)}{1+\|z\|^2}\right)^{\bar{B}(n,d_1,\ldots,d_k)} \leqslant C \cdot \max_i |f_i(z)|$$

for all $z \in \mathbb{C}^n$, where

  (i) $\bar{B}(n,d_1,\ldots,d_k) = \left(\frac{3}{2}\right) B(n,d_1,\ldots,d_k) + \theta$;
  (ii) $B(n,d_1,\ldots,d_k) = d_1 \cdots \cdot d_k$, if $k \leqslant n$; $= d_1 \cdots \cdot d_{n-1} \cdot d_k$, if $k > n$;
  (iii) $j = \#\{i < \min(k,n) - 1 \mid d_i = 2\}$;
  (iv) $\theta = 1$, if $k > n$ and $d_{n-1} = 2$; $= 0$, otherwise.

The exponents $\bar{B}(n,d_1,\ldots,d_k)$ and 2 (in $1+\|z\|^2$) in Theorem are sharp. In fact, the result in [**JKS**] is more general and is valid over an algebraically closed field of any characteristic.

2. Value Distribution Theory

(2.1) Background. By the fundamental theorem of algebra, a d-degree polynomial equation $f(z) = a$ has precisely d solutions (counting multiplicities). To study the analogue for entire or meromorphic functions, R. Nevanlinna created the value distribution theory in 1929. Denote the chordal distance in the Riemann sphere $\mathbb{P}^1$ from w to a by $\|w,a\|$. Then $0 \leqslant \|w,a\| \leqslant 1$. On any complex manifold the exterior derivative splits into $d = \partial + \bar{\partial}$ and twists into $d^c = \frac{i}{4\pi}(\bar{\partial} - \partial)$. The Fubini-Study form Ω is the rotation invariant volume element of the sphere $\mathbb{P}^1$ with total mass 1. If $a \in \mathbb{P}^1$ is fixed and $w \in \mathbb{P}^1 - \{a\}$, then

$$\Omega(w) = dd^c \log \frac{1}{\|w,a\|^2}.$$

A meromorphic function f on $\mathbb{C}$ can be identified with a holomorphic map $f : \mathbb{C} \to \mathbb{P}^1$. For all $0 < s < r$, the characteristic function T_f of f is defined by

$$T_f(r,s) = \int_s^r \int_{\mathbb{C}[t]} f^*(\Omega) \frac{dt}{t} \geqslant 0,$$

where $f^*(\Omega)$ is the pull-back of the form Ω and $\mathbb{C}[t] = \{z \in \mathbb{C} \mid |z| \leqslant t\}$. Take $a \in \mathbb{P}^1$ with $f \not\equiv a$. For $z \in \mathbb{C}$, denote by $\mu_{f,a}(z)$ the a-multiplicity of f at z. Then $\mu_{f,a} : \mathbb{C} \to \mathbb{Z}$ is a non-negative divisor on $\mathbb{C}$ and the function $n_{f,a}$ is defined by

$$n_{f,a}(r) = \sum_{z \in \mathbb{C}[r]} \mu_{f,a}(z), \qquad \text{for all } r \geqslant 0.$$

For all $0 < s < r \in \mathbb{R}$, the counting function $N_{f,a}$ is defined by

$$N_{f,a}(r,s) = \int_s^r n_{f,a}(t) \frac{dt}{t}.$$

The compensation function $m_{f,a}$ of f for a is defined for $r > 0$ by

$$m_{f,a} = \int_{\mathbb{C}\langle r \rangle} \log \frac{1}{\|f,a\|} \sigma \geqslant 0,$$

where $\mathbb{C}\langle r \rangle = \{z \in \mathbb{C} \mid |z| = r\}$. Nevanlinna's First Main Theorem states

$$T_f(r,s) = N_{f,a}(r,s) + m_{f,a}(r) - m_{f,a}(s).$$

Assume that f is not constant. Then $T_f(r,s) \to \infty$ for $r \to \infty$. The defect of f for a is defined by

$$0 \leqslant \delta(f,a) = \liminf_{r \to \infty} \frac{m_{f,a}(r)}{T_f(r,s)} = 1 - \limsup_{r \to \infty} \frac{N_{f,a}(r,s)}{T_f(r,s)} \leqslant 1.$$

If $a \in \mathbb{P}^1 - f(\mathbb{C})$, then $\delta(f,a) = 1$. Nevanlinna's defect relation asserts

(*)
$$\sum_{a \in \mathbb{P}^1} \delta(f,a) \leqslant 2.$$

The value distribution theory for holomorphic curves was developed by H. Cartan, H. Weyl, J. Weyl, L. Ahlfors and W. Stoll. Let V be a complex vector space of dimension $n + 1$. The associated complex projective space $\mathbb{P}(V) = V - \{0\}/\mathbb{C} - \{0\}$. Define $\tau : V \to \mathbb{C}$ by $\tau(\xi) = |\xi|^2$ for $\xi \in V$. The Fubini-Study form Ω on $\mathbb{P}(V)$ is a $(1,1)$-form on $\mathbb{P}(V)$ such that $\mathbb{P}^*(\Omega) = dd^c \log \tau$, where $\mathbb{P}$ is the quotient map. For any hyperplane H in $\mathbb{P}(V)$, there is a unique point $a \in \mathbb{P}(V^*)$ such that $H = \{\mathbb{P}(\xi) \mid \xi \in V, \alpha(\xi) = 0, \mathbb{P}(\alpha) = a\}$. A distance from $x = \mathbb{P}(\xi) \in \mathbb{P}(V)$ to the hyperplane H determined by $a = \mathbb{P}(\alpha) \in \mathbb{P}(V^*)$ is defined by

$$0 \leqslant \|x,a\| = \frac{\langle \xi, \alpha \rangle}{|\xi||\alpha|} \leqslant 1.$$

A holomorphic map $f : \mathbb{C} \to \mathbb{P}(V)$ is called a holomorphic curve. The characteristic function T_f of f is defined by

$$T_f(r,s) = \int_s^r \int_{z \in \mathbb{C}[t]} f^*(\Omega) \frac{dt}{t}.$$

Take $a \in \mathbb{P}(V^*)$ with $\|f,a\| \not\equiv 0$, the compensation function $m_{f,a}$ of f, a is defined by

$$m_{f,a}(r) = \int_{\mathbb{C}\langle r \rangle} \log \frac{1}{\|f,a\|} \sigma \geqslant 0.$$

The counting function $N_{f,a}$ is defined in the same manner as the one defined for one-dimensional case. Also we define the defect $\delta(f,a)$ in the same manner as before.

The holomorphic map $f : \mathbb{C} \to \mathbb{P}(V)$ is said to be linearly non-degenerate if $f(\mathbb{C})$ is not contained in any hyperplane in $\mathbb{P}(V)$. A subset G of $\mathbb{P}(V^*)$ is said to be in general position if every $H \subset G$ with $\#H = p + 1 \leqslant n + 1$ spans a p-dimensional projective linear subspace of $\mathbb{P}(V^*)$. Assume that f is linearly non-degenerate and that $G \neq \emptyset$ is a subset of $\mathbb{P}(V^*)$ such that G is in general position. Then the defect relation holds:

$$(**)\qquad\qquad \sum_{a \in G} \delta(f, a) \leqslant n + 1.$$

H. Cartan first proved $(**)$ in 1933. Later proofs were given by H. and J. Weyl and L. Ahlfors.

(2.2) The Nevanlinna conjecture for moving targets. In 1929, Nevanlinna conjectured that the inequality $(*)$ in (2.1) is still valid if the point $a \in \mathbb{P}^1$ is replaced by a slowly moving meromorphic function $g : \mathbb{C} \to \mathbb{P}^1$, namely,

$$\sum_{g \in T} \delta(f, g) \leqslant 2$$

where T is a finite set of meromorphic functions $\{g \mid T_g(r, s)/T_f(r, s) \to 0,$ as $s \to \infty\}$.

In the case of $\#T = 3$, Nevanlinna easily reduced the conjecture to the defect relation of constant targets. In 1964, C. T. Chung proved the conjecture for entire functions. In 1981, L. Yang showed that $\{g \mid \delta(f, g) > 0, T_g(r, s)/T_f(r, s) \to 0,$ as $r \to \infty\}$ is almost countable if f has finite lower order. In 1986, G. Frank and G. Weissenborn proved the case where T consists of rational functions. In 1986, Steinmetz proved the conjecture.

Consider the holomorphic curve case, it is very natural to ask: does the defect relation $(**)$ still hold when the point $a \in \mathbb{P}(V^*)$ is replaced by a slowly moving holomorphic map $g : \mathbb{C} \to \mathbb{P}(V^*)$? In 1983, S. Mori proved the case $\#\mathcal{F} = n+2$. In 1986, Stoll [Sto2] proved the defect relation with weak bound $n(n + 1)$. In 1989, Ru and Stoll established the theorem with sharp bound.

THEOREM [RS2]. *Let V be a hermitian vector space of dimension $n + 1 > 1$. Let $f : \mathbb{C} \to \mathbb{P}(V)$ be a transcendental holomorphic map. Let $\mathcal{F}$ be a finite set of holomorphic maps $g : \mathbb{C} \to \mathbb{P}(V^*)$, with $\#\mathcal{F} \geqslant n + 1$. Assume that $\mathcal{F}$ is in general position. Assume that f is linearly nondegenerate over $\mathcal{R}_{\mathcal{F}}$, where $\mathcal{R}_{\mathcal{F}}$ is the field generated by all coordinate functions of all $g \in \mathcal{F}$. Assume that $T_g(r, s)/T_f(r, s) \to 0$ as $r \to \infty$ for all $g \in \mathcal{F}$. Then we have*

$$\sum_{g \in \mathcal{F}} \delta(f, g) \leqslant n + 1.$$

The theorem is also valid for holomorphic maps $f : \mathbb{C}^m \to \mathbb{P}(V)$, where m is any positive integer.

When H. Cartan established the defect relation in 1933 for the case of linearly non-degenerate holomorphic curves $f : \mathbb{C} \to \mathbb{P}(V)$, he also conjectured that the defect bound of k-nondegenerate holomorphic curves $f : \mathbb{C} \to \mathbb{P}(V)$ (i.e., $f(\mathbb{C})$ is contained in the k-dimensional subspace of $\mathbb{P}(V)$, but none of lower dimension) should be $2n - k + 1$. In 1983, E. E. Nochka proved the conjecture. Ru and Stoll proved the Cartan's conjecture for the case of slowly moving target hyperplanes as follows.

Let $\mathcal{M}$ be the filed of all meromorphic function on $\mathbb{C}$. Let $\mathbb{K}$ be a subfield of $\mathcal{M}$ with $\mathbb{C} \subset \mathbb{K}$. Let $f : \mathbb{C} \to \mathbb{P}(V)$ be a holomorphic map linearly degenerate over $\mathbb{K}$. The set of all meromorphic maps $g : \mathbb{C} \to \mathbb{P}(V^*)$ such that all functions of g are in $\mathbb{K}$ and $(f, g) = 0$ contains a maximal sequence $g_1, \ldots, g_p$ of linearly independent maps. Put $k = n - p$. Then f is said to be k-flat over $\mathbb{K}$.

THEOREM [RS3]. *Let $\mathcal{F} = \{g_j\}_{j \in Q}$ be a finite family of meromorphic maps $g_j : \mathbb{C}^m \to \mathbb{P}(V^*)$ in general position with $n + 1 \leqslant \#Q < \infty$. Take $k \in \mathbb{Z}[0, n]$. Let $f : \mathbb{C}^m \to \mathbb{P}(V)$ be a transcendental meromorphic map which is k-flat over $\mathcal{R}_{\mathcal{F}}$ such that $(f, g_j) \not\equiv 0$, for each $j \in Q$. Assume that g_j grows slower than f for each $j \in Q$. Then we have*

$$\sum_{j \in Q} \delta(f, g_j) \leqslant 2n - k + 1.$$

(2.3) **Uniqueness problem without multiplicities.** Since the theorems of Pólya, Nevanlinna and Cartan in one complex variable, the uniqueness problem can be divided into two types: the problem with multiplicities and the problem without multiplicities. As the first type problem, Fujimoto proved in 1979 [Fuj1] that if H_j are hyperplanes in $\mathbb{P}^n$ in general position and v_j are divisors on $\mathbb{C}^m$ whose supports have no common irreducible components for $j = 1, \ldots, n + 2$ and $\mathcal{W}$ denotes the set of meromorphic maps $f : \mathbb{C}^m \to \mathbb{P}^n$ with $f^*(H_j) = v_j$ for $j = 1, \ldots, n + 2$, then $\mathcal{W}$ cannot contain more than $n + 1$ algebraically independent maps (when $n = m = 1$ and replace "algebraically independent maps" by "maps", it is the Cartan-Nevanlinna theorem in 1928). Ji proved an analogous result as the second type problem.

THEOREM [Ji2]. *Let $H_1, \ldots, H_k$ be hyperplanes in general position in $\mathbb{P}^n$ with $n \geqslant 2$. Let $A_1, \ldots, A_k$ be pure $(m-1)$-dimensional analytic subsets of $\mathbb{C}^m$ with $\operatorname{codim} A_i \cap A_j \geqslant 2$ whenever $i \neq j$. Suppose $k = 3n + 1$. Then*

   (i) *any linearly non-degenerate meromorphic maps $f, g, h : \mathbb{C}^m \to \mathbb{P}^n$ with $f|A_j = g|A_j = h|A_j$ and with $f^{-1}(H_j) = g^{-1}(H_j) = h^{-1}(H_j) = A_j$ for $j = 1, \ldots, k$ are algebraically dependent (moreover, f, g, h satisfy a polynomial equation of degree 3).*

   (ii) *If in addition $m \geqslant \operatorname{rank} f = \operatorname{rank} g = \operatorname{rank} h = n$, then $k = n + 3$ suffices.*

Previously, S. J. Drouilhet showed [Dr] that in (ii), if $k > n + 3$, then $f = g = h$; and L. M. Smiley showed [Smi] that in (i), if $k > 3n + 1$, then $f = g = h$. Recently, Stoll gave a closely related result [Sto3], which covered the main part of the above theorem, with a different proof.

3. Applications of the Theory of Holomorphic Curves

(3.1) **Gauss map of minimal surfaces immersed in $\mathbb{R}^n$.** Let M be a smooth oriented two-manifold without boundary. Take an immersion $f : M \to$

$\mathbb{R}^n$. The metric on M induced from the standard metric ds_E^2 on $\mathbb{R}^n$ by f is denoted by ds^2. Let Δ be the Laplace-Beltrami operator of (M, ds^2). Make M into a Riemann surface by decreeing that the 1-form $dx + i\,dy$ is of type $(1,0)$, where (x, y) are any isothermal coordinates. We say that f is minimal if $\Delta f = 0$. The Gauss map of f is defined to be

$$G : M \to \mathbb{CP}^{n-1}, \qquad G(z) = \left[\left(\frac{\partial f}{\partial z}\right)\right],$$

where $[(\cdot)]$ denotes the complex line in $\mathbb{C}^n$ through the origin and $(\cdot)$. By the assumption of minimality of M, G is a holomorphic map of M into $\mathbb{CP}^{n-1}$. In 1981, F. Xavier showed that the Gauss map of a nonflat complete minimal surface in $\mathbb{R}^3$ cannot omit seven points of the sphere. In 1988, Fujimoto [**Fuj2**] replaced seven to five, which is sharp. For the $n > 3$ case, Fujimoto [**Fuj3**] proved that the Gauss map of G of a complete minimal surface M in $\mathbb{R}^n$ can omit at most $n(n+1)/2$ hyperplanes in general position in $\mathbb{CP}^{n-1}$, provided G is non-degenerate (i.e., $G(M)$ is not contained in any hyperplane in $\mathbb{CP}^{n-1}$). Fujimoto's non-degenerate condition was removed by Ru as follows.

THEOREM [**Ru1**]. *Let M be a nonflat complete minimal surface immersed in $\mathbb{R}^n$. Then the Gauss map G of M can omit at most $n(n+1)/2$ hyperplanes in $\mathbb{CP}^{n-1}$ located in general position.*

If M is of finite total curvature, we have the following theorem. The case of $n = 3$ was due to R. Osserman, and the non-degenerate case was due to S. S. Chern and R. Osserman [**CO**].

THEOREM [**Ru2**]. *Let M be a nonflat complete minimal surface of finite total curvature immersed in $\mathbb{R}^n$. Then the Gauss map G can omit at most $(n-1)(n+2)/2$ hypersurfaces in $\mathbb{CP}^{n-1}$ located in general position.*

(3.2) Finiteness of integral points in number theory. One of the most important problems in number theory (Diophantine approximation) is the study of integral points or ($\mathbb{K}$)-rational points of a given algebraic variety defined over a number field $\mathbb{K}$. Typically, results come in the form of certain finiteness statements; for instance statements asserting that certain equations have only a finite number of rational (compact case) or integral solutions (affine case). Siegel proved [**Sie**] the following finiteness theorem that $\mathbb{P}^1(\mathbb{K}) - \{3\text{ distinct points}\}$ and $T^1(\mathbb{K}) - \{1\text{ point}\}$ have finitely many integral points in the sense of embedding, where $\mathbb{K}$ is any number field, $\mathbb{P}^1(\mathbb{K})$ is the projective space and $T^1(\mathbb{K})$ is any curve of genus 1. In the compact case, Mordell made the famous conjecture that there are only finitely many rational points on a curve of genus > 1. G. Faltings successfully proved the Mordell conjecture recently. Siegel's theorem was extended to the higher dimensional case. G. Faltings [**Fa2**] proved that if V is an abelian variety and D is a very ample effective divisor, then $V - D$ has only finitely many D-integral points. Ru and Wong extended Siegel's theorem of $\mathbb{P}^1(\mathbb{K}) - \{a_1, a_2, a_3\}$ to the higher dimensional case as below.

THEOREM [**RW**]. *Let* $\mathbb{K}$ *be a number field and* $H_1, \ldots, H_q$ *be a finite set of hyperplanes of* $\mathbb{P}^n(\mathbb{K})$, *assumed to be in general position. Let* $D = \sum_{j=1}^q H_j$. *Then for any integer* $1 \leqslant k \leqslant n$, *the set of* D-*integral points of* $\mathbb{P}^n(\mathbb{K}) - D$ *is contained in a finite union of linear subspaces of* $\mathbb{P}^n(\mathbb{K})$ *of dimension* $k - 1$ *provided* $q > 2n - k + 1$. *In particular, the set of* D-*integral points of* $\mathbb{P}^n(\mathbb{K}) - \{2n + 1 \text{ hyperplanes in general position}\}$ *is finite.*

As an immediate consequence of the theorem, we also have a corollary: let V be a projective variety over $\mathbb{K}$, D a very ample divisor on V. Let $D_1, \ldots, D_q$ be divisors in the linear system $|D|$ such that $E = D_1 + \cdots + D_q$ has, at most, simple normal crossing singularities. If $q > 2N - k + 1$, where $N = \dim \mathcal{L}(D) - 1$ and $1 \leqslant k \leqslant n$, then the set of E-integral points of $V - E$ is contained in the intersection of a finite number of linear subspaces, of dimension $k - 1$, of $\mathbb{P}^N$ with V. In particular, the set of E-integral points of $V - E$ is finite if $q \geqslant 2N + 1$.

Diophantine approximation is a branch of number theory and the Nevanlinna theory is a branch of complex analysis. These two areas of mathematics have been developed totally independently from each other. In the last few years, due to the works of Osgood, Lang, Vojta and others, there appear to be evidence that the theory of Diophantine approximation and the Nevanlinna theory may be somehow related. Vojta has come up with a dictionary for translating results in one theory to the other. The dictionary is essentially formal in the nature and the above results seems to be the first instance where the correspondence works out perfectly.

(3.3) Nevanlinna theorems in push-forward version. We consider polynomial map $f : \mathbb{C}^n \to \mathbb{C}^n$. There is a well-known Jacobian conjecture: if the Jacobian $\det(J_f) = 1$, then f has an inverse of polynomial map. Related to this, Ji proved

THEOREM [**Ji4**]. *Let* f *be a surjective meromorphic map (resp. surjective proper holomorphic map) from* M *onto* N, *where* M *and* N *are* n-*dimensional compact complex manifolds (resp.* n-*dimensional complex manifolds). Let* L *be a holomorphic line bundle with semi-positive curvature and with a nonzero holomorphic section* s. *Then the image of the zero locus of* s *by* f *is an analytic hypersurface on* N.

COROLLARY. *Let* $f : \mathbb{C}^n \to \mathbb{C}^n$ *be a polynomial map with* $\det(J_f) = 1$. *Then*

$$f \text{ has an inverse of polynomial type} \iff \operatorname{supp} F_* D_{z_0} = \operatorname{supp} D_{w_0}$$

where $F : \mathbb{P}^n \to \mathbb{P}^n$ *is the associated rational map, and* D_{z_0} *is the divisor on* $\mathbb{P}^n$ *given by* $\{z_0 = 0\}$, *and* $F_* D_{z_0}$ *is a push-forward current (which is in fact a divisor).*

From the above result, it leads us to pay attention to the push-forward divisor $F_* D_{z_0}$. In order to investigate general push-forward divisors, we establish the Nevanlinna theorems in push-forward version for polynomial maps as follows. The notation is similar to the one in (2.1) (cf. [**Ji5**]).

THEOREM [Ji5].   (i) *(First Main Theorem) Let f, F be as above, and divisor D_g be given by a holomorphic polynomial g. Then*

$$\deg g \cdot T_{*F}(r,s) = N_{*F}(D_g;r,s) + \frac{1}{2}\int_{S(r)} F_*\phi_g\sigma + O(1), \qquad for\ r \gg s.$$

(ii) *(Second Main Theorem) Let $D_1,\ldots,D_q$ be divisors on $\mathbb{P}^n$ so that their supports are in normal crossings. Suppose $D_j = D_{g_j}$, where g_j is a homogeneous polynomial of degree p_j. Denote $D = \sum_j D_j$. Then for $r \gg s$, we have*

$$\left(\sum_{j=1}^q p_j - (n+1)\right) T_{*F}(r,s) \leqslant N_{*F}(D;r,s) + N_{*F}(D_{J_F};r,s) + \log r + O(1)$$

where J_F is the Jacobian.

(iii) *(Defect relation) $\sum_{j=1}^q \delta_{*F}(D_j) \leqslant n+1 + \deg(J_F)$.*

(iv) *Let D be a divisor on $\mathbb{P}^n$ with $\operatorname{codim}(\operatorname{supp} D \cap F(\operatorname{supp}(D_{J_F}))) \geqslant 2$. Let $F^*D = D_g$, and suppose D_{g_1} is smooth and that g_1 divides g, where g, g_1 are homogeneous polynomials. Then*

$$\deg g_1 \leqslant \lambda_F + n+1 + \deg J_F$$

where λ_F is the sheets number of F.

(3.4) Normal families of holomorphic mappings into compact complex manifold. According to Bloch's rule, L. Zalcman [Za] made a principle to determine the normality of meromorphic function families. Zalcman's principle plays an important role in one complex variable. Zalcman's principle was generalized by Ru as follows:

THEOREM [Ru3]. *Let P be a property (i.e., a set) of holomorphic mappings to the compact complex manifold N which satisfies the following conditions.*

(i) *If $\langle f, D \rangle \in P$ and $D' \subset D$, then $\langle f, D' \rangle \in P$;*

(ii) *If $\langle f, D \rangle \in P$ and $\Phi(z) = az + b$, then $\langle f \circ \Phi, \Phi^{-1}(D) \rangle \in P$;*

(iii) *Let $\langle f_n, D_n \rangle \in P$, where $D_1 \subset D_2 \subset \ldots$ and $D = \cup D_n$. If $f_n \to f$ uniformly on compact subsets of D, then $\langle f, D \rangle \in P$;*

(iv) *$\langle f, \mathbb{C} \rangle \in P$ only if f is constant.*

Then for any domain $G \subset \mathbb{C}$, the family of holomorphic maps f such that $\langle f, G \rangle \in P$ is normal.

We have the following Brody's theorem as a corollary.

THEOREM [Bro]. *A complex compact manifold is hyperbolic if and only if it contains no holomorphic curves.*

4. Currents, Line Bundles and Moishezon Manifold

(4.1) Kempf's distortion inequality. Let $X = \mathbb{C}^g/L$ be an abelian variety, where L is a lattice in $\mathbb{C}^g$. Let $\mathcal{L} = \mathcal{L}(\alpha, H)$ be an ample invertible sheaf over X, where (α, H) is the Appel-Humbert data. Consider the Einstein-Hermitian metric $(\ ,\ )_1$ and the Fubini-Study metric $(\ ,\ )_2$ on $\mathcal{L}$. The distortion function $b_{\mathcal{L}(1)}(x)$ is a real analytic function on X defined by $b_{\mathcal{L}(1)}(\ ,\ )_1 = (\ ,\ )_2$. It was proved by G. Kempf that for any X, $\mathcal{L}$ as above, there are positive constant numbers C_1, C_2 such that

$$(\#) \qquad C_1 m^{2g} \leqslant b_{\mathcal{L}(m^2)}(x) \leqslant C_2 m^{2g}$$

for any $x \in X$ and any positive integer m, where $\mathcal{L}(m) = \mathcal{L}^{\otimes m}$. From $(\#)$, it yields

$$(\#\#) \qquad \lim_{m \to \infty} \left(b_{\mathcal{L}(m^2)}(x)\right)^{1/m^2} = 1$$

uniformly for $x \in X$. Kempf conjectured that $(\#\#)$ should be true if m^2 is replaced by m. Roughly speaking, $(\#\#)$ means that the Fubini-Study metric eventually flattens out. Ji proved that Kempf's conjecture is true.

THEOREM [**Ji3**]. *Let X, $\mathcal{L}$, $b_{\mathcal{L}(m)}$ be as above. Then for any $\varepsilon > 0$, there are positive constant numbers $C_1(\varepsilon)$ and $C_2(\varepsilon)$ such that*

$$C_1(\varepsilon) m^{-(3g+2)(2\sqrt{m}+1)-\varepsilon} \leqslant b_{\mathcal{L}(m)} \leqslant C_2(\varepsilon) m^{7g+3+\varepsilon}$$

for any $x \in X$ and any positive integer m. Consequently, one has

$$\lim_{m \to \infty} \left(b_{\mathcal{L}(m)}(x)\right)^{1/m} = 1.$$

i.e., Kempf's conjecture is true.

This result has been generalized by Demailly [**De1**] into any positive line bundle over a projective algebraic manifold. Also G. Tian [**Ti**] independently proved a stronger result.

(4.2) Moishezon manifolds and line bundles. Given a Moishezon manifold M (i.e., the transcendental degree of the meromorphic function filed of M is equal to the dimension of M), it is well-known that there is a disingularization $\pi : \tilde{M} \to M$ such that the manifold $\tilde{M}$ is projective algebraic. Let $\tilde{\omega}$ be a Kähler form on $\tilde{M}$ with $[\tilde{\omega}] \in H^2(M, \mathbb{Z})$, then the push-forward current $\omega = \pi_* \tilde{\omega}$ is a d-closed, strictly positive, integral $(1,1)$-current. Conversely, Shiffman conjectured: let M be a compact complex manifold. Then M is Moishezon if and only if there exists a d-closed, strictly positive integral $(1,1)$-current ω on M. This conjecture is to generalize the Kodaira embedding theorem in terms of currents and has been proved as follows.

THEOREM [Ji7][JS]. *Let X be a compact complex manifold. Then the following statements are equivalent:*

(a) *X is Moishezon;*

(b) *there exists a d-closed, strictly positive, integral $(1,1)$-current ω on X such that $\omega|_{X-S}$ is smooth, where S is a proper analytic subset of X;*

(b') *there exists a holomorphic line bundle L over X with a singular Hermitian metric h such that the curvature current $c_1(L,h)$ is strictly positive and h is smooth on $X - S$, where S is as above;*

(c) *there exists a d-closed, strictly positive integral $(1,1)$-current ω on X;*

(c') *there exists a holomorphic line bundle L over X with a singular Hermitian metric H such that $c_1(L,h)$ is strictly positive;*

(d) *there is a proper analytic subset $S \subset X$ such that $X-S$ admits a complete Kähler-Einstein metric with negative Ricci curvature;*

(e) *there is a proper analytic subset $S \subset X$ such that $X - S$ admits a complete Kähler-Einstein metric with negative Ricci curvature and with finite volume;*

(f) *there is a proper analytic subset $S \subset X$ such that $X-S$ admits a complete Kähler metric g with $\mathrm{Ricci}(g) \leqslant -g$.*

Acknowledgement. Partial financial support for the first author was provided by a University of Houston Initiation Research Grant and by the NSF under grant number DMS-8922760.

REFERENCES

[B]   W. D. Brownawell, *Local Diophantine Nullstellen inequalities*, Journal AMS **1** (1988), 311–322.

[Bro]   R. Brody, *Compact manifolds and hyperbolicity*, Trans. Amer. Math. Soc. **235** (1978), 213–219.

[Car1] H. Cartan, *Sur les zéros des combinations liéaires de p fonctions holomorphes données*, Mathematica(cluj) **7** (1933), 80–103.

[Ch]   W. Chen, *Cartan conjecture: defect relation for meromorphic maps from manifold to projective spaces*, Univ. of Notre Dame Thesis (1987).

[CG]   M. J. Cowen and P. A. Griffiths, *Holomorphic curves and metrics of negative curvature*, J. D'Analyse Math. **29** (1976), 93–153.

[CO]   S. S. Chern and R. Osserman, *Complete minimal surfaces in Euclidean n-spaces*, J. D'Analyse Math. **65** (1967), 15–34.

[De1]   J.-P. Demailly, *Holomorphic Morse inequalities*, Series of Lectures given at AMS Summer Institute held in Santa Cruz, California (July 1989).

[De2]   _____, *Singular Hermitian metrics on positive line bundles*, Proceedings of the conference "complex algebraic varieties" (Bayreuth, April 1990), Lecture Notes in Mathematics, Springer-Verlag (to appear).

[De3]   _____, *Regularization of closed positive currents and intersection theory*, preprint (1991).

[Dr]   S. J. Drouilhet, *A unicity theorem for meromorphic mappings between algebraic varieties*, Trans. Amer. Math. Soc. **265** (1981), 349–358.

[Fa1]   G. Faltings, *Endlichkeitssätze für abelsche Varietäten über Zahlkörpern*, Invent. Math. **73** (1983), 349–366.

[Fa2]   _____, *Diophantine Approximations on Abelian Varieties*, Annals of Math. (to appear).

[Fuj1]   H. Fujimoto, *Remarks to the uniqueness problem of meromorphic maps into* $\mathbb{P}^N(\mathbb{C})$, III, Nagoya Math. J. **58** (1975), 71–85.

[Fuj2]   _____, *On the number of exceptional values of the Gauss map of minimal surfaces*, J. Math. Soc. Japan **49** (1988), 235–247.

[Fuj3]   _____, *Modified defect relations for the Gauss map of minimal surfaces* II, J. Diff. Geom. **31** (1990), 365–385.

[Ji1]   S. Ji, *Some results about common Borel direction of meromorphic functions and their derivatives of arbitrary order*, J. East China Normal Univ. **2** (1981).

[Ji2]   _____, *Uniqueness problem without multiplicities in value distribution theory*, Pacific J. Math. **135** (1988), 323–347.

[Ji3]   _____, *Inequality of distortion functions for invertible sheaves on abelian varieties*, Duke Math. J. **3** (1989), 657–667.

[Ji4]   _____, *Image of analytic hypersurfaces*, Indiana U. of Math. J. **39** (1990), 477–483.

[Ji5]   _____, *Nevanlinna theorems in push-forward version*, Proceedings Symposium on Value Distribution Theory in Several Complex Variables (W. Stoll, ed.), Univ. of Notre Dame Press, 1992, pp. 86–106.

[Ji6]   _____, *Smoothing of currents and Moišezon manifolds*, Proceedings of Symposia in Pure Math. vol. 52, part 2, 1991, pp. 273–281.

[Ji7]   _____, *Currents, metrics and Moishezon manifolds*, Pacific J. Math. (to appear).

[JD1]   S. Ji and C.-J. Dai, *ρ order semilines and the relation among distributions of Borel direction*, J. East China Normal Univ. **2** (1980).

[JD2]   _____, *The growth of functions and common Borel directions*, J. East China Normal Univ. **1** (1981).

[JD3]   _____, *Some results for common Borel directions*, J. East China Normal Univ. **3** (1982).

[JKS]   S. Ji, J. Kollár and B. Shiffman, *A global Lojasiewicz inequality for algebraic varieties*, Trans. Amer. Math. Soc. **329** (1992), 813–818.

[JS]   S. Ji and B. Shiffman, *Properties of compact complex manifolds carrying closed positive currents*, preprint.

[M]   B. Malgrange, *Ideals of Differentiable Functions*, Oxford Univ. Press, 1966.

[Noc]   E. I. Nochka, *On the theory of meromorphic curves*, Soviet Math. Dolk. **27(2)** (1983).

[Ru1]   M. Ru, *On the Gauss map of minimal surfaces immersed in* $\mathbb{R}^n$, J. Diff. Geom. **34** (1991), 411–423.

[Ru2]   _____, *The Gauss map of minimal surfaces of finite total curvature*, Bulletin of the Australian Math. Soc. **44(2)** (1991), 225–232.

[Ru3]   _____, *A general principle for normal families of holomorphic mappings*, preprint.

[Ru4]   _____, *A general theorem on the total number of deficient values for a class of meromorphic functions*, Chinese Ann. of Math. (Ser. A) **9(2)** (1988), 178–187.

[Ru5]   _____, *Some results on the existence of common Borel direction of meromorphic functions*, J. East China Normal Univ. Sci. Ed. No. 1 (1989), 39–50.

[RP]   M. Ru and X. Pang, *On the total number of deficient entire functions*, Chinese Ann. of Math. (Ser. A) **6(4)** (1985), 411–424.

[RS1]   M. Ru and W. Stoll, *Courbes holomorphes evitant huperplans mobiles*, C. R. Acad. Sci. Paris t310, sere I (1990), 45–48.

[RS2]   _____, *The second main theorem for moving targets*, The Journal of Geometric Analysis **1(2)** (1991), 99–138.

[RS3]   _____, *The Cartan conjecture for moving targets*, Proceedings of Symposia in Pure Mathematics vol. 52, part 2, Amer. Math. Soc., 1991, pp. 477–508.

[RS4]   _____, *The Nevanlinna conjecture for moving targets*, Geometrical and Algebraical Aspects in Several Complex Variables (C. A. Berenstein and D. C. Struppas, eds.), Cetraro (Italy), June 1989, pp. 293–308.

[RW]   M. Ru and P. M. Wong, *Integral points of* $\mathbb{P}^n(\mathbb{K}) - \{2n + 1$ *hyperplanes in general position*$\}$, Invent. Math. **106** (1991), 195–216.

[Sch]   W. M. Schmidt, *Diophantine Approximation*, Lecture Notes in Mathematics 785, Springer-Verlag, New York, 1980.

[Sie]   C. L. Siegel, *Approximation algebraischer Zahlen*, Math. Zeitscher. **10** (1921), 173–213.

[Smi]   L. Smiley, *Geometric conditions for unicity of holomorphic curves*, Contemporary Math. vol. 25, Amer. Math. Soc., 1983, pp. 149–154.

[Sto1]   W. Stoll, *Value distribution theory for meromorphic maps*, Aspects of Math. **E7** (1985).

[Sto2]   _____, *An extension of the theorem of Steinmetz-Nevanlinna to holomorphic curves*, Math. Ann. **282** (1988), 185–222.

[Sto3]   _____, *On the propagation of Defendences*, Pacific J. of Math. **139** (1989), 311–337.

[Ti]   G. Tian, *On a set of polarized Kähler metrics on algebraic manifolds*, J. Diff. Geom. **32** (1990), 99–130.

[Vo1]   P. Vojta, *Diophantine Approximations and Value Distribution Theory*, Lecture Notes in Mathematics 1239, Springer-Verlag, 1987.

[Vo2]   _____, *Mordell's conjecture over function fields*, Invent. Math. **98** (1989), 115–138.

[Wo]   P. M. Wong, *Defect relations for maps on parabolic spaces and Kobayashi metric on projective spaces omitting hyperplanes*, Univ. of Notre Dame Thesis (1976).

[Za]   L. Zalcman, *A heuristic principle in complex function theory*, Amer. Math. Monthly **82** (1975), 813–817.

DEPARTMENT OF MATHEMATICS, UNIVERSITY OF HOUSTON, HOUSTON, TX 77204, U. S. A.

DEPARTMENT OF MATHEMATICS, NATIONAL UNIVERSITY OF SINGAPORE, KENT RIDGE CRESCENT, SINGAPORE 0511, REPUBLIC OF SINGAPORE

Contemporary Mathematics
Volume **142**, 1993

Factorization of Meromorphic Functions in Several Complex Variables

BAO QIN LI AND CHUNG-CHUN YANG

1. Introduction

Let $F(z)$ be a meromorphic function in the complex plane $\mathbb{C} = \{z \mid |z| < +\infty\}$. $F(z)$ is said to have a factorization with left factor f and right factor g provided

$$(1.1) \qquad\qquad F(z) = f(g(z)),$$

where f is a meromorphic function in $\mathbb{C}$ and g is an entire function in $\mathbb{C}$ (g may be meromorphic when f is rational). $F(z)$ is said to be prime (pseudo-prime, left-prime, and right-prime) if every factorization of form (1.1) implies that either f is bilinear of g is linear (either f is rational or g is a polynomial, f must be bilinear whenever g is transcendental, g must be linear whenever f is transcendental). The first example of prime functions is $F(z) = e^z + z$ given by Rosenbloom [**R**] who investigated the existence and quantitative estimation of the fixed-point of a composite function of two entire functions. He asserted without proof that $e^z + z$ is prime. The primeness of this function was first proved by Gross [**G1**]. Since then there have been several different proofs which have suggested several directions in the development of the theory. While factorization theory of meromorphic functions of one complex variable has been pursued by numerous authors in various aspects and many related topics may have been intensely studied, (c.f. [**G1, G2, G3, CY, Y1, Y2**]), there is nothing about the theory of several complex variables in the literature.

The main purpose of the present paper is to try to extend and promote the research of factorization theory of meromorphic functions in one complex variable z to several complex variables Z (or ξ). The first problem we encounter here is how to give a proper definition of the primeness of entire or meromorphic functions of several complex variables. Given a meromorphic function $F(Z)$:

1991 *Mathematics Subject Classification.* 32A99, 32A15, 32A22.

Research of the second author is partially supported by an UPGC grant of Hong Kong.

This paper is in final form and no version of it will be submitted for publication elsewhere.

$\mathbb{C}^n \to \mathbb{P}^1 := \mathbb{C} \cup \{\infty\}$, one might consider the following "natural factorization": $F(Z) = f(g)$, where $g : \mathbb{C}^n \to \mathbb{C}^m$ ($m \geqslant 1$ a positive integer) is an entire map and $f : \mathbb{C}^m \to \mathbb{P}^1$ is a meromorphic function, and define the corresponding concept of primeness, accordingly. However, it is easy to see that such a definition is not compatible with the theory in one variable as is shown by the example $F(z) = e^z + z$. In fact, $F(z)$ is prime in $\mathbb{C}$, but $F(z) = f(g(z))$ with $g = (e^z, 1 + ze^{-z})$, an entire map from $\mathbb{C}$ to $\mathbb{C}^2$ and $f = \xi_1\xi_2$, an entire function from $\mathbb{C}^2$ to $\mathbb{C}$. Both of f and g are non-linear. Hence the usual prime function is not prime any more in the above sense. This observation shows that it is reasonable just to consider the factorization of "functions", without any "maps" involved. We thus pose the following general definition for the primeness of a meromorphic function of several complex variables.

DEFINITION 1.1. Let $F(Z)$ be a meromorphic function in $\mathbb{C}^n$. $F(Z)$ is said to have a factorization with left factor f and right factor g provided

$$(1.2) \qquad\qquad F(Z) = f(g(Z)),$$

where f is a meromorphic function from $\mathbb{C}$ to $\mathbb{P}^1$ and g is an entire function from $\mathbb{C}^n$ to $\mathbb{C}$ (g may be a meromorphic function from $\mathbb{C}^n$ to $\mathbb{P}^1$ when f is a rational function from $\mathbb{C}$ to $\mathbb{P}^1$). $F(Z)$ is said to be prime in $\mathbb{C}^n$ (pseudo-prime, left-prime, and right-prime) if every factorization of form (1.2) implies that either f is bilinear in $\mathbb{C}$ or g is linear in $\mathbb{C}^n$ (either f is rational in $\mathbb{C}$ or g is a polynomial in $\mathbb{C}^n$, f must be bilinear in $\mathbb{C}$ whenever g is transcendental in $\mathbb{C}^n$, g must be linear in $\mathbb{C}^n$ whenever f is transcendental in $\mathbb{C}$).

As mentioned above, $F(z) = e^z + z$ is the first example ever exhibited to be prime in $\mathbb{C}$. Thus it becomes very natural for us to test whether its "extension" in $\mathbb{C}^n$ is prime. That is, can we prove that $F(\xi) = e^{\xi_1 + \xi_2 + \cdots + \xi_n} + a_1\xi_1 + \cdots + a_n\xi_n$ (a_i's are complex numbers, not all zero) is prime in $\mathbb{C}^n$? An affirmative answer will be given as a direct consequence of our main result Theorem 3.1.

We divide the paper into four parts. The first part is the introduction. In Section 2, we recall some basic notations and properties of the growth and divisor of a meromorphic function in $\mathbb{C}^n$ which are needed in our proofs. In Section 3, we give a criterion for the primeness of a meromorphic function in $\mathbb{C}^n$ (Theorem 3.1) which will enable us to show, as a corollary, that the above function $F(\xi) = e^{\sum_{i=1}^{n} \xi_i} + \sum_{i=1}^{n} a_i\xi_i$ is prime. In the final section, we compare the relation between the primeness of a meromorphic function $F(\xi)$ in $\mathbb{C}^n$ and its restrictions $F(c_1, c_2, \ldots, \xi_i, \ldots, c_n)$ for all complex ξ_i-axis.

2. The growth of a meromorphic function

This section will provide a brief account of notations and basic facts of value distribution theory of several complex variables that are needed in the proofs of our results. Let $f(\xi)$ be a meromorphic function in $\mathbb{C}^n$; $d\sigma_r$ be the positive element of volume on the sphere

$$S_n(r) = \{\xi \in \mathbb{C}^n \mid |\xi| = r\},$$

Considered as a real $(2n-1)$-dimensional C^∞ manifold, oriented to the exterior of the ball

$$B_n(r) = \{\xi \in \mathbb{C}^n \mid |\xi| < r\}.$$

Also let

$$V_n(r) = \frac{2\pi^n r^{2n-1}}{(n-1)!}$$

be the surface area of $S_n(r)$ and

$$\omega_n(\xi_1, \xi_2, \ldots, \xi_n) = \frac{1}{(n-1)!} \left\{ \sum_{j=1}^{n} \frac{i}{2} d\xi_j \wedge d\bar{\xi}_j \right\}^{n-1},$$

a C^∞ complex exterior differential form of bidegree $(n-1, n-1)$ on $\mathbb{C}^n$.

Then the characteristic of f is defined by

$$T_f(r, s) = \int_s^r \frac{A_f(t)}{t} dt \qquad \text{for } r \geqslant s > 0,$$

where

$$A_f(t) := \frac{(n-1)!}{\pi^n t^{2n-2}} \int_{B_n(t)} \frac{i}{2(1+|f|^2)^2} df \wedge d\bar{f} \wedge \omega_n.$$

Since $A_f(t)$ is a nonnegative, increasing, continuous function on $\mathbb{R}^+ := [0, +\infty)$ (see e.g. [**K, S**]), $T_f(\cdot, s)$ is a nonnegative increasing C^1 function on $[s, +\infty)$. The order of f is defined by

$$\rho(f) = \limsup_{r \to \infty} \frac{\log^+ T_f(r, s)}{\log r}, \qquad \text{for some } s > 0.$$

For a meromorphic function f in $\mathbb{C}$, the above characteristic differs from the standard Nevanlinna characteristic $T(r, f)$ by a constant (see e.g. [**Ku**]). There

$$T(r, f) = N(r, f) + m(r, f),$$

with

$$N(r, f) = \int_0^r \frac{n(t, f) - n(0, f)}{t} dt + n(0, r) \log r,$$

$$m(r, f) := \frac{1}{2\pi} \int_0^{2\pi} \log^+ |f(re^{i\theta})| d\theta,$$

and $n(r, f) :=$ number of poles of f in the disc $|z| \leqslant r$ (see e.g. [**H**]).

For a function $f \not\equiv 0$ holomorphic on an open connected neighborhood of ξ in $\mathbb{C}^n$, we have the following series

$$f(\zeta) = \sum_{j=\nu}^{\infty} P_i(\zeta - \xi)$$

which converges uniformly on some neighborhood of ξ and represents f on this neighborhood. Here P_i is a homogeneous polynomial of degree i and $P_\nu \not\equiv 0$. The nonnegative integer ν, uniquely determined by f and ξ, is called the zero-divisor of f at ξ and is denoted by $\mathrm{div}_f(\xi)$ or $\mathrm{div}_f^0(\xi)$.

If f is a non-constant meromorphic function in $\mathbb{C}^n$, then for each $\xi \in \mathbb{C}^n$ there exist an open connected neighborhood U_ξ of ξ in $\mathbb{C}^n$ and two holomorphic functions $g \not\equiv 0$ and $h \not\equiv 0$ on U_ξ, coprime at ξ (i.e., the germs of g and h are coprime in the local ring of germs of holomorphic functions at ξ), such that $hf = g$ on U_ξ. The nonnegative integers $\mathrm{div}_f^0(\xi) := \mathrm{div}_g(\xi)$ and $\mathrm{div}_f^\infty(\xi) := \mathrm{div}_h(\xi)$, uniquely determined by f and ξ, are called respectively, the zero-divisor and the pole-divisor of f. $\mathrm{div}_f(\xi) := \mathrm{div}_f^0(\xi) - \mathrm{div}_f^\infty(\xi)$ is called the divisor of f. $\mathrm{div}_f^\alpha(\xi) := \mathrm{div}_{g-\alpha h}(\xi)$ is called the α-divisor of f if $\alpha \in \mathbb{C}$. Clearly, the function f is holomorphic on $\mathbb{C}^n$ if and only if $\mathrm{div}_f(\xi) \geqslant 0$ on $\mathbb{C}^n$ and f is holomorphic without zeros on $\mathbb{C}^n$ if and only if $\mathrm{div}_f(\xi) \equiv 0$ on $\mathbb{C}^n$.

If f is an entire function in $\mathbb{C}^n$ and $\mathrm{div}_f \equiv \mathrm{div}_P$ for a polynomial P in $\mathbb{C}^n$, then div_f is called an algebraic divisor and $Z(f) := \{\xi \in \mathbb{C}^n \mid f(\xi) = 0\}$ is called an algebraic variety. Otherwise, $Z(f)$ is said to be transcendental. For an algebraic variety $Z(f)$, if $Z(f) = Z(P)$ for a polynomial P of degree $\leqslant 1$ in $\mathbb{C}^n$, then $Z(f)$ is called an algebraic variety of degree $\leqslant 1$, If such a polynomial P does not exist, $Z(f)$ is called an algebraic variety of degree higher than 1. For any meromorphic function f in $\mathbb{C}^n$, there are two entire functions g and h such that $hf = g$ on $\mathbb{C}^n$ and g and h are coprime at every point of $\mathbb{C}^n$ (see e.g. [**Ho**]). Then $Z(f)$ is called algebraic if $Z(g)$ is algebraic and $P(f) := Z(h)$ is called algebraic if $Z(h)$ is algebraic. We also use the notation $Z(f, g) := \{\xi \in \mathbb{C}^n \mid f(\xi) = g(\xi) = 0\}$ for any meromorphic functions f and g in $\mathbb{C}^n$, and the notation $D_u f := \dfrac{\partial f}{\partial \xi_1} u_1 + \cdots + \dfrac{\partial f}{\partial \xi_n} u_n$, the directional derivative of f along a unit vector $u \in S^{2n-1} := S_n(1)$. Now define the differential form

$$\eta = \frac{1}{4\pi} dd^c |\xi|^2,$$

where $d = \partial + \bar{\partial}$ and $d^c = \dfrac{\partial - \bar{\partial}}{i}$. For a meromorphic function f in $\mathbb{C}^n$ and $\alpha \in \mathbb{P}^1$, the counting function of the α-divisor of f is defined by

$$n_f^\alpha(r) = r^{2-2n} \int_{|\xi| \leqslant r, \, \xi \in \mathrm{supp}(\mathrm{div}_f^\alpha)} \mathrm{div}_f^\alpha \, \eta^{n-1}$$

for $n > 1$ and

$$n_f^\alpha(r) = \sum_{|z| \leqslant r} \mathrm{div}_f^\alpha(z)$$

for $n = 1$. The valence function of the α-divisor is defined by

$$N_f^\alpha(r, s) = \int_r^s n_f^\alpha(t) \frac{dt}{t}, \qquad \text{for } r \geqslant s > 0.$$

The compensation function of f for $\alpha \in \mathbb{P}^1$ is defined by

$$m_f^\alpha(r) = \frac{1}{V_n(r)} \int_{S_n(r)} \log \frac{1}{\|f, \alpha\|} d\sigma_r,$$

where $\|f, \alpha\|$ is the chordal distance between f and α on the Riemann surface $\mathbb{P}^1$.

One has the following first main theorem (see e.g. [**V**, **S**]):

$$T_f(r, s) = N_f^\alpha(r, s) + m_f^\alpha(r) - m_f^\alpha(s)$$

for any $\alpha \in \mathbb{P}^1$. This implies that

$$0 \leqslant \delta_f(\alpha) := \liminf_{r \to \infty} \frac{m_f^\alpha(r)}{T_f(r, s)} = 1 - \limsup_{r \to \infty} \frac{N_f^\alpha(r, s)}{T_f(r, s)} \leqslant 1,$$

where $\delta_f(\alpha)$ is called the defect of f for α.

3. Criteria for the primeness of meromorphic functions

THEOREM 3.1. *Let $F(\xi)$ be an entire function of finite order in $\mathbb{C}^n$ and $u \in S^{2n-1}$ be a unit vector. If $Z(D_u F)$ is a transcendental variety and for any $a \in \mathbb{C}$, $Z(F-a, D_u F)$ does not contain any transcendental variety or polynomial variety of degree higher than 1. Then F is prime.*

PROOF. First we prove that F is pseudo-prime. Assume, to the contrary, that $F(\xi) = f(g(\xi))$, where $g : \mathbb{C}^n \to \mathbb{C}$ is a transcendental entire function and $f : \mathbb{C} \to \mathbb{P}^1$ is a transcendental meromorphic function.

We consider the restriction $F_\zeta(z) : \mathbb{C} \to \mathbb{C}$ of $F(\xi)$ to $\zeta \in S^{2n-1}$ defined by $F_\zeta(z) := F(\zeta z)$. By [**Ku** Th. 3.9], for any meromorphic function G in $\mathbb{C}^n$ with $\mathrm{div}_G(0) = 0$ and $s > 0$ there are constants A, B and C in $\mathbb{R}^+$ such that $T_{G_\xi}(r, s) \leqslant A T_G(Br, s) + C$ for $r > 0$ and $\xi \in S^{2n-1}$. Therefore one can deduce immediately, for $\zeta \in S^{2n-1}$,

$$\limsup_{r \to \infty} \frac{\log^+ T(r, F_\zeta)}{\log r} < +\infty$$

by the assumption that F is of finite order. That is, each restriction F_ζ to $\zeta \in S^{2n-1}$ is of finite order. Since $g(\xi)$ is transcendental, we have the following series expansion

$$g(\xi) = \sum_{j \geqslant 0}^{\infty} P_j(\xi),$$

where P_j is a homogeneous polynomial of degree j and there are infinitely many j's such that $P_j \not\equiv 0$. Denote by J the set of such j's. For $j \in J$ the set

$$Z_j := \{\xi \in \mathbb{C}^n \mid P_j(\xi) = 0\}$$

is a thin variety and so has $2n$-dim zero measure (see e.g. [**Ra**]). Therefore $V := \cup_{j \in J} Z_j$ also has $2n$-dim zero measure. Take $\zeta \in \mathbb{C}^n$ such that $\zeta \neq 0$ and $\zeta \notin V$ and denote $\hat{\zeta} = \frac{\zeta}{|\zeta|}$. Then for such j,

$$P_j(\hat{\zeta}) = P_j \left(\frac{\zeta}{|\zeta|} \right) = \frac{1}{|\zeta|^j} P_j(\zeta) \neq 0.$$

This implies that $g_{\hat{\zeta}}(z) := g(\hat{\zeta}z) = \sum_{j \geqslant 0}^{\infty} z^j P_j(\hat{\zeta})$ is a transcendental entire function of z in $\mathbb{C}$. Notice that $F_{\hat{\zeta}}(z) = f(g_{\hat{\zeta}}(z))$ and recall that $F_{\hat{\zeta}}$ is of finite order. We obtain that f must be of order zero by the theorem in [**EF**]: if f is a meromorphic function of positive order in $\mathbb{C}$ and g is a transcendental entire function in $\mathbb{C}$ then $f(g(z))$ is of infinite order. Also, for any meromorphic function h in $\mathbb{C}$, h and its derivative h' have the same order (see e.g. [**H**]). We deduce that f' is of order zero. We assert that f' has at most one pole. Otherwise, it follows from Picard theorem (see e.g. [**H**]) for one complex variable that, for some pole z_0 of f, $g_{\zeta_0}(z) = z_0$ would have infinitely many roots which are all poles of $F(\zeta)$, a contradiction. Thus by the Hadamard factorization for meromorphic function in $\mathbb{C}$ (see e.g. [**T**]), we see that f' must have infinitely many zeros, $\{z_m\}_{m=1}^{\infty}$, say. We claim that for all $m \in \mathbb{N}$ (the set of natural integers) with at most two exceptions, the set $Z(g - z_m)$ is a transcendental variety. In fact, if for some m, $Z(g - z_m)$ is algebraic, then there exists a polynomial $P_m(\zeta)$ in $\mathbb{C}^n$ such that $Z(g - z_m) = Z(P_m)$. Remember for any polynomial $P(\zeta)$ in $\mathbb{C}^n$, $T_P(r, s) = O(\log r)$ and for any transcendental meromorphic function $h(\xi)$ in $\mathbb{C}^n$, $\log r = o\{T_h(r, s)\}$ as $r \to \infty$ (see e.g. [**G, S**]). Therefore $N_g^{z_m}(r, s) = N_{P_m}^0(r, s) \leqslant T_{P_m}(r, s) = O(\log r)$ and

$$\log r = o\{T_g(r, s)\}$$

as $r \to \infty$. This implies that the defect

$$\delta_g(z_m) = 1 - \limsup_{r \to \infty} \frac{N_g^{z_m}(r, s)}{T_g(r, s)} = 1.$$

However, by the defect relation (see [**V**] or [**S**]) for meromorphic functions in $\mathbb{C}^n$,

$$\sum_{m=1}^{\infty} \delta_g(z_m) \leqslant 2.$$

Therefore, the above claim holds. On the other hand, it is easy to see that $Z(g - z_n) \subset Z(F - f(z_n), D_u F)$. We arrive at a contradiction to the hypothesis that for any $a \in \mathbb{C}$, $Z(F - a, D_u F)$ does not contain any transcendental variety. We have thus proved the pseudo-primeness of F.

Now we prove that F is left-prime. By the above result, we may assume that $F(\xi) = P(g(\xi))$, where $P(z) : \mathbb{C} \to \mathbb{P}^1$ is a rational function of order $d \geqslant 1$ and $g(\xi) : \mathbb{C}^n \to \mathbb{P}^1$ is a meromorphic function. It suffices to prove that $d = 1$. We discuss two cases.

Case 1: $P(\infty) = \infty$. In this case $g(\xi)$ must be an entire function in $\mathbb{C}^n$ since otherwise $F(\xi)$ would have poles. We assert that $P'(z)$ cannot have any zeros in $\mathbb{C}$. If, to the contrary, $P'(z_0) = 0$ for some $z_0 \in \mathbb{C}$, then $\operatorname{div}_{g-z_0}$ must be an algebraic divisor since otherwise we would have that the transcendental variety $Z(g-z_0)$ belongs to $Z(F - P(z_0), D_u F)$, a contradiction to the hypothesis of the theorem. Thus we can find a polynomial $R(\xi)$ in $\mathbb{C}^n$ such that $\operatorname{div}_{g-z_0} = \operatorname{div}_R$ in $\mathbb{C}^n$. This means that $\dfrac{g(\xi) - z_0}{R(\xi)}$ is an entire function without any zeros in $\mathbb{C}^n$. We then have that

$$(3.1) \qquad\qquad g(\xi) - z_0 = R(\xi)e^{S(\xi)},$$

where $S(\xi)$ is an entire function in $\mathbb{C}^n$. For any $\zeta \in S^{2n-1}$, $F_\zeta(Z) = P(g_\zeta(Z))$. Since F is of finite order, its restriction F_ζ to ζ must be of finite order (see the above proof of the pseudo-primeness). But by [**BJV**], for any meromorphic function $h(z)$ and rational function $Q(z)$ of degree q in $\mathbb{C}$, $T(r, Q(h)) = qT(r, h) + O(\log r)$ as $r \to \infty$. Therefore $g_\zeta(z)$ is of finite order for each $\zeta \in S^{2n-1}$. By (3.1), $g_\zeta(z) - z_0 = R_\zeta(z)e^{S_\zeta(z)}$. Hence for each $\zeta \in \mathbb{C}^n$ with $\zeta \notin Z(R)$, which is a thin variety, we must have that $S_{\hat\zeta}(Z)$ is a polynomial in $\mathbb{C}$, where $\hat\zeta = \dfrac{\zeta}{|\zeta|}$. This shows that in (3.1), $S(\xi)$ turns out to be a polynomial in $\mathbb{C}^n$. Otherwise, by the same proof as in the pseudo-primeness, for all $\zeta \in \mathbb{C}^n$ except for a set of $2n$-dim zero measure, $S_{\hat\zeta}(Z)$ is transcendental, a contradiction to the above. Obviously, by (3.1)

$$D_u g(\xi) = (D_u R(\xi) + R(\xi)D_u S(\xi))e^{S(\xi)}.$$

Since $R(\xi)$ and $S(\xi)$ are all polynomials in $\mathbb{C}^n$, we see that

$$\operatorname{div}_{D_u g} = \operatorname{div}_{D_u R + R D_u S} + \operatorname{div}_{e^S} = \operatorname{div}_{D_u R + R D_u S},$$

an algebraic divisor. By the hypothesis, $Z(D_u F)$ is a transcendental variety and $D_u F = P'(g)D_u g$. Therefore there must exist a $z_1 \in \mathbb{C}$ such that $P'(z_1) = 0$ and $Z(g - z_1)$ is a transcendental variety. This is a contradiction since $Z(g - z_1) \subset Z(F - P(z_1), D_u F)$. It thus follows that $P'(z)$ cannot have any zeros in $\mathbb{C}$. Set

$$P(z) = A\frac{q_1(z)}{q_2(z)} = A\frac{(z - a_1)^{n_1}(z - a_2)^{n_2}\ldots(z - a_N)^{n_N}}{(z - b_1)^{m_1}(z - b_2)^{m_2}\ldots(z - b_M)^{m_M}},$$

where A is a non-zero constant, $q_1(z)$ and $q_2(z)$ are two coprime polynomials in $\mathbb{C}$. Since $P(\infty) = \infty$, $d = \sum_{j=1}^N n_j > e := \sum_{j=1}^M m_j \geqslant 0$. It is easy to see that

$$P'(z) = A\frac{q_1'(z)\prod\limits_{j=1}^M (z - b_j) - q_1(z)\sum\limits_{j=1}^M \left(m_j \prod\limits_{i=1,i\neq j}^M (z - b_i)\right)}{\prod\limits_{j=1}^M (z - b_j)^{m_j+1}}$$

$$= A \frac{q(z)}{\displaystyle\prod_{j=1}^{M}(z - b_j)^{m_j+1}}$$

with the obvious definition of $q(z)$. Since $q_1(b_j) \neq 0$ for $1 \leqslant j \leqslant M$, $q(b_j) \neq 0$ for $1 \leqslant j \leqslant M$. Thus $q(z)$ and $\prod_{j=1}^{M}(z - b_j)^{m_j+1}$ are coprime. However the leading term of $q(z)$ is $dz^{d-1+M} - (m_1 + m_2 + \cdots + m_M)z^{d+M-1} = (d - e)z^{d+M-1}$. Recall that $P'(z)$ and so $q(z)$ do not have any zero in $\mathbb{C}$. We thus deduce that $d + M - 1 = 0$ and so that $d = 1$ and $e = 0$ by the fact that $d > e \geqslant M \geqslant 0$.

Case 2: $P(\infty) \neq \infty$. In this case, $P(z)$ must have a pole, w_0 say. Hence

$$P(z) = \frac{1}{(z - w_0)^m}q_3(z) = \left\{ w^m q_3 \left(w_0 + \frac{1}{w} \right) \right\} := q_4(w)$$

where m is a positive integer, $q_3(z)$ is a rational function in $\mathbb{C}$ with $q_3(w_0) \neq 0$, and $w = \dfrac{1}{z - w_0}$ is a Möbius transformation. Clearly $q_4(w)$ is a rational function of w and $q_4(\infty) = \infty$. Notice that

$$F(\xi) = P(g(\xi)) = q_4 \left(\frac{1}{g(\xi) - w_0} \right).$$

By the result of case (1), we must have that

$$q_4(w) = aw + b$$

for some constants a, b with $a \neq 0$ and so that

$$P(z) = q_4(w) = q_4 \left(\frac{1}{z - w_0} \right) = a\frac{1}{z - w_0} + b$$

is a rational function of degree 1, i.e. $d = 1$. This completes the proof of the left-primeness of F.

Next, it remains to prove the right-primeness. By the pseudo-primeness of F, we may assume that

$$F(\xi) = f(Q(\xi)),$$

where f is a transcendental meromorphic function from $\mathbb{C}$ to $\mathbb{P}^1$ and $Q(\xi)$ is a polynomial of degree ρ ($\geqslant 1$) in $\mathbb{C}^n$. Clearly

$$D_u F = f'(Q)D_u Q.$$

By the hypothesis that $Z(D_u F)$ is transcendental, $f'(z)$ must have infinitely many zeros, $\{a_m\}_{m=1}^{\infty}$, say. Since for any $m \in \mathbb{N}$,

$$Z(Q - a_m) \subset Z(F - f(a_m), D_u F)$$

and $Z(F - f(a_m), D_u F)$ does not contain any transcendental variety and algebraic variety of degree higher than 1 by the hypothesis, we deduce that

$$Z(Q - a_m) = Z(Q_m)$$

for a polynomial Q_m of degree $\leqslant 1$ in $\mathbb{C}^n$. By writing $Q - a_m$ into the product of irreducible functions, it is easy to see that $q - a_m$ must assume the following form

$$(3.2) \qquad Q - a_m = \left(c_m^{(0)} + c_m^{(1)} \xi_1 + c_m^{(2)} \xi_2 + \cdots + c_m^{(n)} \xi_n \right)^\rho,$$

where $c_m^{(0)}, \ldots, c_m^{(n)}$ are constants and ξ_j is the j-th coordinate of ξ. Also we can write $Q(\xi)$ into a Weierstrass pseudo polynomial in some variable, ξ_1 say, such that, in view of (3.2),

$$Q(\xi) = Q(\xi_1, \xi_2, \ldots, \xi_n) = \alpha_\rho \xi_1^\rho + \alpha_{\rho-1}(\xi_2, \ldots, \xi_n) \xi_1^{\rho-1} + \cdots + \alpha_0(\xi_2, \ldots, \xi_n),$$

where $\alpha_\rho \neq 0$ is a constant, and $\alpha_{\rho-1}, \ldots, \alpha_0$ are polynomials of $\xi_2, \ldots, \xi_n$. Therefore by letting $\xi_2 = 0, \xi_3 = 0, \ldots, \xi_n = 0$, we get a polynomial $Q^*(\xi_1) := Q(\xi_1, 0, 0, \ldots, 0)$ of degree ρ in ξ_1. It follows that for large m, $Q^*(\xi_1) - a_m$ has ρ distinct zeros in the plane. However by (3.2), $Q^*(\xi_1) - a_m = (c_m^{(0)} + c_m^{(1)} \xi_1)^\rho$ has only one zero of multiplicity ρ. Therefore it follows that $\rho = 1$. This shows that F is right-prime.

Summarizing the above results, we have showed that F is both left-prime and right-prime. Therefore F must be prime. The proof of the theorem is thus complete. $\square$

Observing that in the proof of the left-primeness of Theorem 3.1, we did not use the condition that "$Z(F - a, D_u F)$ does not contain any polynomial variety of degree higher than 1", we thus have the following proposition for the left-primeness.

PROPOSITION 3.2. *Let $F(\xi)$ and $u \in S^{2n-1}$ be the same as in Theorem 3.1. If $Z(D_u F)$ is a transcendental variety and for any $a \in \mathbb{C}$, $Z(F - a, D_u F)$ does not contain any transcendental variety, then F is left-prime.*

REMARK 3.3. It is natural to ask whether or not F is prime under the hypotheses of proposition 3.2. A negative answer is given by the following counter example. Consider, in $\mathbb{C}^n$, the function

$$F(\xi) = \cos(\xi_1 + \xi_2 + \cdots + \xi_n) + (\xi_1 + \xi_2 + \cdots + \xi_n)^2.$$

Clearly, $F(\xi)$ is an entire function of finite order. Let

$$u = (1, 0, \ldots, 0).$$

Then

$$(3.3) \qquad D_u F = \frac{\partial F}{\partial \xi_1} = -\sin(\xi_1 + \xi_2 + \cdots + \xi_n) + 2(\xi_1 + \xi_2 + \cdots + \xi_n).$$

We assert that $Z(D_u F)$ is a transcendental variety. Otherwise there would be two polynomials $P_1(\xi)$ and $P_2(\xi)$ in $\mathbb{C}^n$ such that

$$D_u F = P_1(\xi) e^{P_2(\xi)}$$

and so that

$$D_u^{(3)} F := D_u(D_u(D_u F))$$

$$= \left\{ \frac{\partial}{\partial \xi_1} \left(\frac{\partial P_1}{\partial \xi_1} + P_1 \frac{\partial P_2}{\partial \xi_1} \right) + \left(\frac{\partial P_1}{\partial \xi_1} + P_1 \frac{\partial P_2}{\partial \xi_1} \right) \frac{\partial P_2}{\partial \xi_1} \right\} e^{P_2}.$$

This means that $\mathrm{div}_{D_u^{(3)} F}$ is a polynomial divisor. However by (3.3),

$$\mathrm{div}_{D_u^{(3)} F} = \sin(\xi_1 + \xi_2 + \cdots + \xi_n)$$

and thus $\mathrm{div}_{D_u^{(3)} F} = \mathrm{div}_{\sin(\xi_1 + \xi_2 + \cdots + \xi_n)}$, a non-polynomial divisor. We obtain a contradiction. Next, for any $a \in \mathbb{C}$, it is easy to check that

$$Z(F - a, D_u F) \subset Z(Q(Z)),$$

where $Q(Z) := \{a - (\xi_1 + \cdots + \xi_n)^2\}^2 + 4(\xi_1 + \cdots + \xi_n) - 1$ is a polynomial in $\mathbb{C}^n$. Thus $Z(F - a, D_u F)$ does not contain any transcendental variety. But

$$F(Z) = (\cos \sqrt{Z} + Z) \circ (\xi_1 + \cdots + \xi_n)^2$$

is not prime in $\mathbb{C}^n$. This example shows that in Theorem 3.1, one cannot drop the condition: "$Z(F - a, D_u F)$ does not contain any polynomial variety of degree higher than 1".

COROLLARY 3.4. $F(\xi) = e^{(\xi_1 + \xi_2 + \cdots + \xi_n)} + a_1 \xi_1 + a_2 \xi_2 + \cdots + a_n \xi_n$ is prime, where a_i's are complex numbers not all zero.

PROOF. It is no loss of generality to assume that $a_1 \neq 0$. Let $u = (1, 0, \ldots, 0)$. Then

$$D_u F = e^{\xi_1 + \xi_2 + \cdots + \xi_n} + a_1.$$

Clearly F is of finite order, $Z(D_u F)$ is a transcendental variety and for any $a \in \mathbb{C}$,

$$Z(D_u F, F - a) = Z(Q(\xi)),$$

where $Q(\xi) = a_1 \xi_1 + a_2 \xi_2 + \cdots + a_n \xi_n - (a - a_1)$ is a polynomial of degree 1. By Theorem 3.1, $F(Z)$ is prime. $\square$

There are a large number of functions satisfying the hypotheses of Theorem 3.1 and thus being prime, for example, it is easy to verify that $F_1(\xi) = \sin(\xi_1 + \xi_2 + \cdots + \xi_n) + a_1 \xi_1 + \cdots + a_n \xi_n$ and $F_2(\xi) = \cos(\xi_1 + \xi_2 + \cdots + \xi_n) + a_1 \xi_1 + \cdots + a_n \xi_n$ are all prime, where a_i's are complex numbers not all zero.

Letting $n = 1$ in proposition 3.2 and Theorem 3.1, we have the following result for the factorization of one complex variable, which is interesting in itself.

COROLLARY 3.5. *Let $F(z)$ be an entire function of finite order in the complex plane. If $F'(z)$ has infinitely many zeros and for any $a \in \mathbb{C}$, $Z(F - a, F')$ contains finitely many points, then F is left-prime. Moreover, if $Z(F - a, F')$ contains at most one point, then F is prime.*

REMARK 3.6. As pointed out in Section 1, $F(z) = e^z + z$ is a fundamental prime function in $\mathbb{C}$. Several different proofs have been found in the literature.

Now the primeness of this function is a direct consequence of Corollary 3.4 or Corollary 3.5. We also remark here that for an entire function $F(z)$, one can define the concepts of E-prime, E-pseudo-prime, E-left-prime and E-right-prime. F is called E-prime (E-pseudo-prime, E-left-prime, E-right-prime) if F is prime (pseudo-prime, left-prime, right-prime) when only entire factors f and g are considered in the factorization of form (1.1) or (1.2). Obviously, that F is prime (pseudo-prime, left-prime, right-prime) implies that F is E-prime (E-pseudo-prime, E-left-prime, E-right-prime). But the converse is not true even in the case when $n = 1$ (For example, the function $e^{-mz} + e^{-nz}$ is E-prime but not prime, where $m, n \in \mathbb{N}$). As a direct consequence of Corollary 3.5, we also have the corresponding criteria for the E-primeness of entire functions in $\mathbb{C}$. In particular, one obtains Ozawa's result: under the hypotheses of Corollary 3.5, F is E-left-prime (see [O]).

we conclude this section by pointing out that in Theorem 3.1, the function F can also be meromorphic in $\mathbb{C}^n$ provided that $P(F)$ is algebraic. That is, we have the following

THEOREM 3.7. *Let $F(\xi)$ be a meromorphic function of finite order in $\mathbb{C}^n$ and $u \in S^{2n-1}$ be a unit vector. If $P(F)$ is an algebraic variety, $Z(D_u F)$ is a transcendental variety and for any $a \in \mathbb{C}$, $Z(F - a, D_u F)$ does not contain any transcendental variety or polynomial variety of degree higher than 1. Then F is prime.*

The proo f is a simple modification of the one of Theorem 3.1. We thus omit the details here.

4. The comparison between the primeness of $F(\xi)$ and $F(c_1, c_2, \ldots, \xi_i, \ldots, c_n)$

Let $F(\xi)$ be a meromorphic function in $\mathbb{C}^n$. We consider the relation between the primeness of $F(\xi)$ and its restriction $F(c_1, c_2, \ldots, \xi_i, \ldots, c_n)$ to each complex ξ_i-axis. It is natural to ask whether or not the primeness of $F(\xi)$ is equivalent to the primeness of $F(c_1, c_2, \ldots, \xi_i, \ldots, c_n)$ for $1 \leqslant i \leqslant n$ and any fixed constants c_j $(1 \leqslant j \leqslant n, j \neq i)$. The answer is negative.

Let us consider the entire function

$$F(\xi_1, \xi_2, \ldots, \xi_n) = \left(1 + \prod_{j=1}^{n} \xi_j\right) e^{\prod_{j=1}^{n} \xi_j} + \prod_{j=1}^{n} \xi_j$$

in $\mathbb{C}^n$. Clearly

$$F(\xi_1, \xi_2, \ldots, \xi_n) = ((1 + z)e^z + z) \circ \left(\prod_{j=1}^{n} \xi_j\right)$$

is not prime in $\mathbb{C}^n$. However, for any $1 \leqslant i \leqslant n$ and fixed constants c_j $(1 \leqslant j \leqslant n,$

$j \neq i$),

$$F(c_1, c_2, \ldots, \xi_i, \ldots, c_n) = \left(1 + \left(\prod_{j=1, j \neq i}^{n} c_j\right)\xi_i\right)e^{(\prod_{j=1, j \neq i}^{n} c_j)\xi_i} + \left(\prod_{j=1, j \neq i}^{n} c_j\right)\xi_i$$

$$= (1 + c\xi_i)e^{c\xi_i} + c\xi_i \qquad \left(c := \prod_{j=1, j \neq i}^{n} c_j\right)$$

is a prime function of ξ_i (see e.g. [G]). This example shows that the primeness of $F(c_1, c_2, \ldots, \xi_i, \ldots, c_n)$ for $1 \leqslant i \leqslant n$ does not guarantee that $F(\xi)$ is prime in $\mathbb{C}^n$

In the opposite direction, the primeness of $F(\xi)$ also does not guarantee $F(c_1, c_2, \ldots, \xi_i, \ldots, c_n)$ is prime as shown by the following counter example. Consider the function

$$F(\xi) = F(\xi_1, \xi_2, \ldots, \xi_n) = e^{\sum_{j=1}^{n} \xi_j} + \xi_1.$$

By Corollary 3.4, $F(\xi)$ is prime in $\mathbb{C}^n$. However for any fixed constants c_j ($1 \leqslant j \leqslant n-1$), $F(c_1, c_2, \ldots, c_{n-1}, \xi_n)$, the restriction of $F(\xi)$ to the variable ξ_n, is not prime in ξ_n since

$$F(c_1, c_2, \ldots, c_{n-1}, \xi_n) = e^{\sum_{j=1}^{n-1} c_j + \xi_n} + c_1 = f(g),$$

where $f(Z) = Z^2 + c_1$ and $g(\xi_n) = e^{\frac{1}{2}(\sum_{j=1}^{n-1} c_j + \xi_n)}$.

In the positive direction, we have the following theorem which shows that by adding some extra conditions one can get the primeness of $F(\xi)$ from the primeness of all restrictions of $F(\xi)$. For the sake of simpleness, we restrict to two complex variables.

THEOREM 4.1. *Let $F(\xi)$ be a meromorphic function in $\mathbb{C}^2$. If $F(c, z)$, $F(z, c)$ and $F(sz, tz)$ are all prime in $\mathbb{C}$, where c, s, t are any fixed complex numbers with $st \neq 0$, then $F(\xi)$ is prime in $\mathbb{C}^2$.*

PROOF. Let $F(\xi) = f(g(\xi))$ be a factorization of form (1.2). Then

$$F(c, z) = f(g(c, z)) \quad \text{and} \quad F(z, c) = f(g(z, c)).$$

Since $F(c, z)$ is prime, we have that either f is bilinear or g is linear. If f is bilinear, then there is nothing to prove. We may assume that f is not linear. Then $g(c, z)$ and $g(z, c)$ must be linear. We thus have

$$g(c, \xi_2) = a_1(c)\xi_2 + a_2(c),$$

and

$$g(\xi_1, c) = b_1(c)\xi_1 + b_2(c),$$

where a_1, a_2, b_1, b_2 are meromorphic functions in $\mathbb{C}$.

It follows that

(4.1) $$g(\xi_1, \xi_2) = a_1(\xi_1)\xi_2 + a_2(\xi_1) = b_1(\xi_2)\xi_1 + b_2(\xi_2).$$

In (4.1), letting $\xi_2 = 0$, we have

$$a_2(\xi_1) = b_1(0)\xi_1 + b_2(0)$$

and letting $\xi_2 = 1$, we have

(4.2) $$a_1(\xi_1) + a_2(\xi_1) = b_1(1)\xi_1 + b_2(1).$$

Therefore,

$$\begin{aligned} a_1(\xi_1) &= b_1(1)\xi_1 + b_2(1) - a_2(\xi_1) \\ &= b_1(1)\xi_1 + b_2(1) - b_1(0)\xi_1 - b_2(0) := a\xi_1 + b, \end{aligned}$$

where $a = b_1(1) - b_1(0)$ and $b = b_2(1) - b_2(0)$. Then by (4.2),

$$\begin{aligned} a_2(\xi_1) &= b_1(1)\xi_1 + b_2(1) - a_1(\xi_1) \\ &= b_1(1)\xi_1 + b_2(1) - a\xi_1 - b := c\xi_1 + d, \end{aligned}$$

where $c = b_1(1) - a$ and $d = b_2(1) - b$. Now (4.1) yields that

(4.3) $$\begin{aligned} g(\xi_1, \xi_2) &= (a\xi_1 + b)\xi_2 + c\xi_1 + d \\ &= a\xi_1\xi_2 + b\xi_2 + c\xi_1 + d. \end{aligned}$$

On the other hand, $F(sz, tz)$ is a prime function of z. Hence $g(sz, tz)$ must be linear. This implies that $a = 0$ in (4.3). We then obtain that $g(\xi_1, \xi_2) = b\xi_2 + c\xi_1 + d$ is a linear function in $\mathbb{C}^2$.

The proof of Theorem 4.1 is thus complete. □

Theorem 4.1 enables us to decide whether or not some functions are prime in $\mathbb{C}^n$ by using known results in $\mathbb{C}$. For example, one can immediately see that $e^{\xi_1} + e^{\xi_2} + \xi_1 + \xi_2$ is a prime function in $\mathbb{C}^2$.

REFERENCES

[BG] I. N. Baker and F. Gross, *Further results on factorization of entire functions*, Entire Functions and Related Parts of Analysis, (Proc. Symp. Pure Math., La Jolla, CA 1966), American Mathematical Society, Providence, RI, 1968, pp. 30–35.

[BJV] W. Bergweiler, G. Jank and L. Volkman, *Wachstumsverhalten Zusammengesetzer Funktionen*, Result in Math. **7** (1984), 35–53.

[CY] C. T. Chuang and C. C. Yang, *Theory of Fix Points and Factorization of Meromorphic Functions*, Mathematical Monograph Series, Peking University Press, 1986.

[EF] A. Edrie and W. H. J. Fucks, *On the zeros of $f(g(z))$ where f and g are entire functions*, J. Analyse Math. **12** (1964), 243–255.

[G] P. A. Griffiths, *Entire holomorphic mapping in one and several variables*, Princeton University Press, 1976.

[G1] F. Gross, *On factorization of meromorphic functions*, Trans. Amer. Math. Soc. **131** (1968), 215–222.

[G2] _____, *On factorization of meromorphic functions which are periodic mod g*, Indian Journ. Pure Appl. Math. **2** (1971), 561–571.

[G3] _____, *Factorization of meromorphic functions*, U. S. Government Printing Office, Washington D. C., 1972.

[H] W. K. Hayman, *Meromorphic Functions*, Oxford Math. Monographs, Clarendon Press, Oxford, 1964.

[Ho] L. Hörmander, *An Introduction to Complex Analysis in Several Variables*, Van Nostrand, Princeton, NJ, 1966.

[K]   H. Kneser, *Zur Theorie der gebrochenen Funktionen mehrerer Veränderlichen*, Jber. Deutsch Math. Verein **48** (1938), 1–28.

[Ku]   R. O. Kujala, *Functions of finite ∧-type in several complex variables*, Trans. Amer. Math. Soc. **161** (1971), 327–358.

[O]   M. Ozawa, *On certain criteria for the left-primeness of entire functions*, Kodai Math. Sem. Rep. **6** (1975), 304–317.

[Ra]   R. M. Range, *Holomorphic Functions and Integral Representations in Several Complex Variables*, Springer-Verlag, New York, 1986.

[R]   P. C. Rosenbloom, *The fix-points of entire functions*, Medd. Lunds Univ. Mat. Sem. Suppl. Bd. M. Riesz (1952), 186–192.

[S]   W. Stoll, *Introduction to Value Distribution Theory of Meromorphic Maps*, Lecture Notes in Math. No. 950, Springer-Verlag, New York, 1982, pp. 210–359.

[T]   E. C. Titchmarsh, *The Theory of Functions*, 2nd ed., Oxford Univ. Press, London, 1939.

[V]   A. Vitter, *The lemma of the logarithmic derivative in several complex variables*, Duke Math. J. **44** (1977), 89–104.

[Y1]   C. C. Yang, *On the dependence of zeros of an entire function and its factorization*, J. Math. Anal. Appl. **35** (1971), 374–380.

[Y2]   _____, *Some aspects of factorization theory—a survey*, Complex Variables **13** (1989), 133–142.

DEPARTMENT OF MATHEMATICS, UNIVERSITY OF MARYLAND, COLLEGE PARK, MD 20742, U. S. A.

DEPARTMENT OF MATHEMATICS, THE HONG KONG UNIVERSITY OF SCIENCE AND TECHNOLOGY, KOWLOON, HONG KONG

Contemporary Mathematics
Volume **142**, 1993

Some Results on Singular Integrals and Function Spaces in Several Complex Variables

SHI JI-HUAI

ABSTRACT. In this paper, we shall mainly introduce some works on singular integrals and function spaces in several complex variables, which were done by Gong Sheng, the author and their colleagues in recent years.

§1. Singular Integrals

1.1 Boundary Value of Cauchy Integrals in the Unit Ball

In one complex variable, the importance of Cauchy integrals is well known. After Professor Hua [**17**] determined the Cauchy kernel of classical domains in 1958, the study of the Cauchy integrals of classical domains has become possible.

By B we denote the unit ball of $\mathbb{C}^n$, S its boundary. The Cauchy kernel of B is

$$\omega_{2n-1}^{-1}(1 - z\bar{u}')^{-n}\dot{u},$$

where $u \in S$, $z \in B$, $\omega_{2n-1} = \dfrac{2\pi^n}{\Gamma(n)}$ the volume of S, $\dot{u}$ denote the volume element of S.

If f is integrable on S, the Cauchy integral of f

$$F(z) = \frac{1}{\omega_{2n-1}} \int_S \frac{f(u)\dot{u}}{(1 - z\bar{u}')^n}$$

determines a holomorphic function in B. Gong Sheng and Shi Ji-Huai studied the limit of $F(z)$ as z tends to S in 1982.

1991 *Mathematics Subject Classification.* 32A35, 32A37, 32A40, 32-02.

Research supported by the National Science Foundation of China.

This paper is in final form and no version of it will be submitted for publication elsewhere.

Let f be integrable on S and $v \in S$, they defined the Cauchy principal value of F at v as follows:

$$\text{P. V.} \frac{1}{\omega_{2n-1}} \int_{S(\gamma)} \frac{f(u)}{(1 - v\bar{u}')^n} \dot{u} = \lim_{\varepsilon \to 0} \frac{1}{\omega_{2n-1}} \int_{\Sigma_\varepsilon} \frac{f(u)}{(1 - v\bar{u}')^n} \dot{u},$$

where

$$\Sigma_\varepsilon = \left\{ u \in S : \alpha^2 (1 - \operatorname{Re} v\bar{u}')^2 + \beta^2 (\operatorname{Im} v\bar{u}')^2 \geq \varepsilon^2 \right\}$$

and $\alpha \geq 0$, $\beta \geq 0$, $\alpha + \beta \neq 0$, $\gamma = \dfrac{\alpha}{\beta}$.

THEOREM 1.1.1 [12, 13]. *Let* $f \in \operatorname{Lip} p$, $0 < p \leq 1$, *then*

$$\text{P. V.} \frac{1}{\omega_{2n-1}} \int_{S(\gamma)} \frac{f(u)}{(1 - v\bar{u}')^n} \dot{u}$$

exists and equals

$$\frac{1}{\omega_{2n-1}} \int_S \frac{f(u) - f(v)}{(1 - v\bar{u}')^n} \dot{u} + \left\{ 1 - \frac{1}{2} \left(\frac{2\beta}{\alpha + \beta} \right)^{n-1} \right\} f(v).$$

Here $f \in \operatorname{Lip} p$ *means that* $f : S \to \mathbb{C}$ *satisfies*

$$|f(u) - f(v)| = O(|u - v|^p), \qquad 0 < p \leq 1$$

for any $u, v \in S$.

Using this result, they obtained the generalized Plemelj's formula.

THEOREM 1.1.2 [12, 13]. *Let* $f \in \operatorname{Lip} p$, $0 < p \leq 1$, $v \in S$, *then*

$$\text{K-lim}_{z \to v} \frac{1}{\omega_{2n-1}} \int_S \frac{f(u)\dot{u}}{(1 - z\bar{u}')^n} = \text{P. V.} \frac{1}{\omega_{2n-1}} \int_{S(\gamma)} \frac{f(u)\dot{u}}{(1 - v\bar{u}')^n}$$

$$+ \frac{1}{2} \left(\frac{2\beta}{\alpha + \beta} \right)^{n-1} f(v).$$

For the definition of K-lim see [13].

If $\alpha = \beta = 1$, Σ_ε is equivalent to

$$|1 - v\bar{u}'| \geq \varepsilon.$$

In this case, theorems 1.1.1 and 1.1.2 have been obtained by Gong Sheng and Sun Jiguang in 1965 [8].

Another case $\beta = 0$, must be noted. In this case, the Plemelj's formula becomes very simple

$$(1.1.1) \qquad \text{K-lim}_{z \to v} \frac{1}{\omega_{2n-1}} \int_S \frac{f(u)\dot{u}}{(1 - z\bar{u}')^n} = \text{P. V.} \frac{1}{\omega_{2n-1}} \int_{S(\infty)} \frac{f(u)\dot{u}}{(1 - v\bar{u}')^n}.$$

In other words, one can find a manner to define the principal value of Cauchy integral, such that this principal value is just the limit value of the Cauchy integral as z approaches $v \in S$. It is impossible in one variable that the limit value of the Cauchy integral can be represented by a certain principal value of the Cauchy integral.

If we regard that the deleted neighborhood around the boundary point v

$$|1 - v\bar{u}'|^2 = (1 - \operatorname{Re} v\bar{u})^2 + (\operatorname{Im} v\bar{u}')^2 \geqslant \varepsilon^2$$

as a "disk", then the deleted neighborhood Σ_ε is an "ellipse". Now we consider the situation that the deleted neighborhood around the boundary point v is a "rectangle". Denote

$$D(v, \varepsilon) = \{u \in S : 1 - \operatorname{Re} v\bar{u}' < \alpha\varepsilon, |\operatorname{Im} v\bar{u}'| < \beta\varepsilon\}$$

and $D^*(v, \varepsilon) = S - D(v, \varepsilon)$, where $\alpha > 0$ and $\beta > 0$. In this case, they defined the Cauchy principal value as follows:

$$\mathrm{P.\,V.} \frac{1}{\omega_{2n-1}} \int_{S(R)} \frac{f(u)\dot{u}}{(1 - v\bar{u}')^n} = \lim_{\varepsilon \to 0} \int_{D^*(v,\varepsilon)} \frac{f(u)\dot{u}}{(1 - v\bar{u}')^n}.$$

THEOREM 1.1.3 [12, 13]. *Let* $f \in \operatorname{Lip} p$, $0 < p \leqslant 1$, *then*

$$\mathrm{P.\,V.} \frac{1}{\omega_{2n-1}} \int_{S(R)} \frac{f(u)\dot{u}}{(1 - v\bar{u}')^n}$$

exists and equals

$$\frac{1}{\omega_{2n-1}} \int_S \frac{f(u) - f(v)}{(1 - v\bar{u}')^n} \dot{u} + (1 - b)f(v),$$

where

$$b = \frac{2^{n-1}}{\pi} \left\{ \frac{\pi}{2} - \int_0^{\arctan \frac{\beta}{\alpha}} \cos^{n-2} t \frac{\sin(n-1)t}{\sin t} dt \right\}.$$

THEOREM 1.1.4 [12, 13]. *Let* $f \in \operatorname{Lip} p$, $0 < p \leqslant 1$, $v \in S$, *then*

$$\mathrm{K\text{-}lim}_{z \to v} \frac{1}{\omega_{2n-1}} \int_S \frac{f(u)\dot{u}}{(1 - z\bar{u}')^n} = \mathrm{P.\,V.} \frac{1}{\omega_{2n-1}} \int_{S(R)} \frac{f(u)\dot{u}}{(1 - v\bar{u}')^n} + bf(v).$$

If $\beta = \infty$, then $b = 0$, we can obtain (1.1.1) from theorem 1.1.4.

1.2 The derivatives of Cauchy integral in the unit ball

In one complex variable, if L is a smooth closed curve, the derivative of the following Cauchy integral

$$\frac{1}{2\pi i} \int_L \frac{f(\zeta)}{\zeta - z} d\zeta$$

is also a Cauchy intergal. In several complex variables, the derivative of Cauchy integral is no longer a Cauchy integral.

In 1984, Gong Sheng and Shi Ji-Huai studied the limit value of the derivatives of Cauchy integral in B. Let

$$F(z) = \frac{1}{\omega_{2n-1}} \int_S \frac{f(u)\dot{u}}{(1 - z\bar{u}')^n}$$

and

$$P_j = \{v = (v_1, \ldots, v_j, \ldots, v_n) : v \in S, v_j = 0\},$$
$$Q_j = \{v = (v_1, \ldots, v_j, \ldots, v_n) : v \in S, v_j \neq 0\}.$$

They found that for the points of P_j the limit value of $\dfrac{\partial F(z)}{\partial z_j}$ can be represented by the Cauchy principal value and it must be represented by the Hadamard principal value for the points of Q_j. By the uniformity of these two kinds of principal values, it can be represented uniformly by Hadamard principal value.

THEOREM 1.2.1 [33]. *Let* $f : \bar{B} \to \mathbb{C}$, $\dfrac{\partial f}{\partial z_k}$, $\dfrac{\partial f}{\partial \bar{z}_k}$ ($k = 1, \ldots n$) *are in* Lip α_k *and* Lip β_k *on* $\bar{B}$ *respectively. Then*

$$\text{K-}\lim_{z \to v} \frac{\partial F}{\partial z_j}(z) = \frac{n}{\omega_{2n-1}} \int_S \frac{\bar{u}_j f(u)}{(1 - v\bar{u}')^{n+1}} \dot{u} + \frac{1}{2} \left(\frac{2\beta}{\alpha + \beta} \right)^n \frac{\partial f}{\partial u_j}(v)$$

holds for $v \in P_j$. *Here the integral*

$$\frac{n}{\omega_{2n-1}} \int_S \frac{\bar{u}_j f(u)}{(1 - v\bar{u}')^{n+1}} \dot{u}$$

is defined as

$$\lim_{\varepsilon \to 0} \frac{n}{\omega_{2n-1}} \int_{\sigma_\varepsilon(v)} \frac{\bar{u}_j f(u)}{(1 - v\bar{u}')^{n+1}} \dot{u},$$

and

$$\sigma_\varepsilon(v) = \{u \in S : \alpha^2(1 - |v\bar{u}|^2)^2 + 4\beta^2(\operatorname{Im} v\bar{u})^2 > \varepsilon^2\}.$$

THEOREM 1.2.2 [33]. *Let* $f : \bar{B} \to \mathbb{C}$ *and* $f \in C^3(\bar{B})$, *then*

$$\lim_{\varepsilon \to 0} \frac{1}{\omega_{2n-1}} \int_{\sigma_\varepsilon(v)} \frac{\bar{u}_k[f(u) - f(v)]}{(1 - v\bar{u}')^{n+1}} \dot{u} \qquad (k = 1, \ldots, n)$$

exists for any $v \in S$.

The above limit is called the Hadamard principal value of singular integral

$$\frac{1}{\omega_{2n-1}} \int_S \frac{\bar{u}_k f(u)}{(1 - v\bar{u}')^{n+1}} \dot{u}$$

and denote it by

$$\text{P.} \frac{n}{\omega_{2n-1}} \int_S \frac{\bar{u}_k f(u)}{(1 - v\bar{u}')^{n+1}} = \lim_{\varepsilon \to 0} \frac{n}{\omega_{2n-1}} \int_{\sigma_\varepsilon(v)} \frac{\bar{u}_k[f(u) - f(v)]}{(1 - v\bar{u}')^{n+1}} \dot{u},$$

$k = 1, \ldots, n$.

THEOREM 1.2.3 [33]. *Let $f : \bar{B} \to \mathbb{C}$ and $f \in C^3(\bar{B})$, then*

$$\text{K-}\lim_{\varepsilon \to 0} \frac{\partial F(z)}{\partial z} = \text{P.} \frac{n}{\omega_{2n-1}} \int_S \frac{\bar{u} f(u)}{(1 - v\bar{u}')^{n+1}} \dot{u} + \frac{1}{2} \left(\frac{2\beta}{\alpha + \beta} \right)^{n-1}$$

$$\left\{ \frac{\partial f(v)}{\partial u} \left[\frac{2\beta}{\alpha + \beta} I - n \frac{\beta - \alpha}{\beta + \alpha} v' \bar{v} \right] - n \frac{\partial f(u)}{\partial \bar{u}} \bar{v}' \bar{v} \right.$$

$$\left. + \frac{2\beta}{\alpha + \beta} \bar{v} \left[\text{tr} \left(\frac{\partial^2 f(v)}{\partial u \partial \bar{u}} \right) - v \frac{\partial^2 f(v)}{\partial u \partial \bar{u}} \bar{v}' \right] \right\}$$

holds for any $v \in S$. Here $\dfrac{\partial F}{\partial z} = \left(\dfrac{\partial F}{\partial z_1}, \ldots, \dfrac{\partial F}{\partial z_n} \right)$, $\dfrac{\partial f}{\partial u} = \left(\dfrac{\partial f}{\partial u_1}, \ldots, \dfrac{\partial f}{\partial u_n} \right)$, $\dfrac{\partial f}{\partial \bar{u}} = \left(\dfrac{\partial f}{\partial \bar{u}_1}, \ldots, \dfrac{\partial f}{\partial \bar{u}_n} \right)$,

$$\frac{\partial^2 f}{\partial u \partial \bar{u}} = \begin{bmatrix} \dfrac{\partial^2 f}{\partial u_1 \partial \bar{u}_1} & \cdots & \dfrac{\partial^2 f}{\partial u_1 \partial \bar{u}_n} \\ \vdots & \ddots & \vdots \\ \dfrac{\partial^2 f}{\partial u_n \partial \bar{u}_1} & \cdots & \dfrac{\partial^2 f}{\partial u_n \partial \bar{u}_n} \end{bmatrix}.$$

Taking $\alpha = \beta$ in Theorem 1.2.3, we have

$$\text{K-}\lim_{\varepsilon \to 0} \frac{\partial F}{\partial z} = \text{P.} \frac{n}{\omega_{2n-1}} \int_S \frac{\bar{u} f(u)}{(1 - v\bar{u}')^{n+1}} \dot{u}$$

$$+ \frac{1}{2} \left\{ \frac{\partial f(v)}{\partial u} - n \frac{\partial f(v)}{\partial \bar{u}} \bar{v}' \bar{v} + \bar{v} \left[\text{tr} \left(\frac{\partial^2 f(v)}{\partial u \partial \bar{u}} \right) - v \frac{\partial^2 f(v)}{\partial u \partial \bar{u}} \bar{v}' \right] \right\}.$$

1.3 Cauchy integrals of classical domains

The first kind of classical domain $\mathcal{R}_I(m, n)$ is the set of $m \times n$ complex matrices Z satisfying

$$I - Z\bar{Z}' > 0.$$

By $L_I^{(k)}$ we denote the set of $Z \in \overline{\mathcal{R}_I(m, n)}$ with

$$\text{rank}(I - Z\bar{Z}') = k.$$

In particular, $L_I^{(0)}$ is the Silov boundary of $\mathcal{R}_I(m, n)$.

The Cauchy Kernel of $\mathcal{R}_I(m, n)$ is

$$H_I(Z, \bar{U}) = \det(I - Z\bar{U}')^{-n},$$

where $Z \in \mathcal{R}_I(m, n)$ and $U \in L_I(m, n)$, which was obtained by Hua [17] in 1958.

Gong, Sun and Shi studied the following Cauchy integral of $\mathcal{R}_I(m, n)$

(1.3.1) $$F(Z) = (V(L_I(m, n)))^{-1} \int_{L_I(m,n)} \phi(U) \det(I - Z\bar{U}')^{-n} \dot{U},$$

where $V(L_I(m, n))$ is the volume of $L_I(m, n)$, $\dot{U}$ the volume element of $L_I(m, n)$.

They obtained

THEOREM 1.3.1 [9, 10, 13]. *Suppose $\phi(U)$ is in* Lip p *$(0 < p \leqslant 1)$ on L_I.* *Then when* $Z = U_0' \begin{bmatrix} \rho & 0 \\ 0 & Z_1 \end{bmatrix} V_0 \in \mathcal{R}_I(m, n)$ *approaches* $Q = U_0' \begin{bmatrix} 1 & 0 \\ 0 & Z_1 \end{bmatrix} V_0 \in L_I(m, n)$, *the limit value of the Cauchy integral (1.3.1) exists and equals*

$$\text{P. V. } V(L_I(m, n))^{-1} \int_{L_I(m,n)(\gamma)} \phi(U) \det(I - Q\bar{U}')^{-n} \dot{U}$$

$$+ \frac{1}{2} \det(I - Z_1 \bar{Z}_1')^{-1} V(L_I(m-1, n-1))^{-1}$$

$$\times \int_{L_I(m-1,n-1)} \phi \left(U_0' \begin{bmatrix} 1 & 0 \\ 0 & U_1 \end{bmatrix} V_0 \right)$$

$$\times \det(I - U_1 \bar{Z}_1') \det(I - Z_1 \bar{U}_1')^{-(n-1)} \dot{U}_1.$$

Here the definition of principal value is

$$\lim_{\varepsilon \to 0} V(L_I(m, n))^{-1} \int_{D(\varepsilon)} \phi(U) \det(I - Q\bar{U}')^{-n} \dot{U}$$

and

$$D(\varepsilon) = \left\{ U \in L_I(m, n) \mid \alpha^2 (\operatorname{Re} \det S_1)^2 + \beta^2 (\operatorname{Im} \det S_1)^2 \geqslant \varepsilon^2 \right\},$$

$$S_1 = I - Q\bar{U}' - U\bar{T}_0' + Q\bar{T}_0', \qquad T_0 = U_0' \begin{bmatrix} 0 & 0 \\ 0 & Z_1 \end{bmatrix} V_0, \qquad \gamma = \frac{\alpha}{\beta}.$$

In particular, when $\beta = 0$, the limit value of Cauchy integral may be represented by a certain Cauchy principal value:

$$\lim_{\varepsilon \to 0} F(Z) = \text{P. V. } V(L_I(m, n))^{-1} \int_{L_I(m,n)(\infty)} \phi(U) \det(I - Q\bar{U}')^{-n} \dot{U}.$$

We say that ϕ is in Lip p on $L_I(m, n)$, if

$$|\phi(U) - \phi(V)| = O \left((\operatorname{tr}(U - V)(\overline{U - V})')^{\frac{p}{2}} \right)$$

holds for any $U, V \in L_I(m, n)$.

The fourth kind of classical domain or Lie sphere hyperbolic space $\mathcal{R}_{IV}$ is the set of complex vectors $z = (z_1, \ldots, z_N)$ satisfying

$$\begin{cases} 1 + |zA_0 z'|^2 - z\bar{z}' > 0 \\ 1 - |zA_0 z'| > 0 \end{cases}$$

and $A_0 = \frac{1}{2} \begin{bmatrix} 0 & 1 \\ -1 & 0 \end{bmatrix} + \frac{1}{2} I^{(N-2)}$.

Denote the boundary and Silov boundary of $\mathcal{R}_{IV}$ by b_{IV} and L_{IV} respectively. The Cauchy kernel of $\mathcal{R}_{IV}$ is

$$H_{IV}(z, \zeta) = (1 + zA_0 z' \overline{\zeta A_0 \zeta'} - z\bar{\zeta}')^{-\frac{N}{2}},$$

where $z \in \mathcal{R}_{IV}$ and $\zeta \in L_{IV}$, and the Cauchy integral of $\mathcal{R}_{IV}$ is

$$(1.3.2) \qquad F(z) = V(L_{IV})^{-1} \int_{L_{IV}} H_{IV}(z, \zeta)\phi(\zeta)\dot{\zeta}.$$

On the Cauchy integral (1.3.2), Gong, Sun and Shi obtained the following result.

THEOREM 1.3.2 [9, 10, 13]. *Suppose* $\phi(z)$ *is in* Lip p $(0 < p < 1)$ *on* L_{IV}. *If* z *approaches* $\eta_0 = e^{i\theta_0}(t, 1, 0, \ldots, 0)P_0^{-1}\Gamma P_0 \in b_{IV} - L_{IV}$ *satisfying the following condition*

$$|1 + \eta_0 A_0 \eta_0' \overline{\zeta A_0 \zeta'} - \eta_0 \bar{\zeta}'| \, |1 + z_0 A_0 z_0 \overline{\zeta A_0 \zeta'} - z_0 \bar{\zeta}'|^{-1} \leqslant M \qquad (M = const.)$$

then

$$\lim_{z \to \eta_0} F(z) = \mathrm{P.\,V.}\, V(L_{IV})^{-1} \int_{L_{IV}(v)} H_{IV}(\eta_0, \zeta)\phi(\zeta)\dot{\zeta} + \frac{1}{2}\left(\frac{2\beta}{\alpha+\beta}\right)^{\frac{N}{2}-1}$$

$$\times \frac{1}{2\pi}\int_0^{2\pi} \phi\left(e^{i\theta_0}\left(e^{i\theta_0}\frac{1+te^{-i\theta}}{1+te^{i\theta}}, 1, 0, \ldots, 0\right)P_0^{-1}\Gamma P_0\right)\frac{d\theta}{(1+te^{i\theta})^{\frac{N}{2}}},$$

where the principal value is defined as

$$\mathrm{P.\,V.}\, V(L_{IV})^{-1}\int_{L_{IV}(v)} H_{IV}(\eta_0, \zeta)\phi(\zeta)\dot{\zeta}$$

$$= \lim_{\varepsilon \to 0} V(L_{IV})^{-1}\int_{G(\varepsilon)} H_{IV}(\eta_0, \zeta)\phi(\zeta)\dot{\zeta}$$

and

$$G(\varepsilon) = \left\{\zeta \in L_{IV} \mid \alpha^2[\mathrm{Re}(1 + \eta_0 A_0 \eta_0'\overline{\zeta A_0 \zeta'} - \eta_0\bar{\zeta}')(1 - \bar{z}_0\zeta')]^2 \right.$$
$$\left. + \beta^2[\mathrm{Im}(1 + \eta_0 A_0 \eta_0'\overline{\zeta A_0 \zeta'} - \eta_0\bar{\zeta}')(1 - \bar{z}_0\zeta')]^2 \geqslant \varepsilon^2\right\}.$$

In particular, if $\beta = 0$ and $N > 2$, then

$$\lim_{z \to \eta_0} F(z) = \mathrm{P.\,V.}\int_{L_{IV}(\infty)} H_{IV}(\eta_0, \zeta)\phi(\zeta)\dot{\zeta}.$$

The other results of Cauchy integrals of classical domains see [7, 9, 10, 13].

1.4 Henkin integrals on strictly pseudoconvex domains

In several complex variables, there is no perfect analogue of Cauchy Kernel, what come closest to it are certain Cauchy-Fantappie (C-F) kernels

$$K(w, z) = \frac{c_n}{g^n}\omega \wedge dz_1 \wedge \cdots \wedge dz_n,$$

where

$$c_n = (-1)^{\frac{n(n-1)}{2}}(n-1)!(2\pi i)^{-n},$$

$$g(w,z) = \sum_{i=1}^{n}(z_i - w_i)g_i(w,z),$$

$$\omega = g_1\bar{\partial}g_2 \wedge \cdots \wedge \bar{\partial}g_n + \cdots + (-1)^{n-1}g_n \wedge \bar{\partial}g_1 \wedge \cdots \wedge \bar{\partial}g_{n-1}.$$

These kernels are different from the Cauchy kernel of one complex variable, they depend on the domain Ω. But they can be constructed for a wide class of Ω—smooth, bounded strictly pseudoconvex domains. Henkin, Ramirez, Stein and Kerzman have constructed some important C-F kernels which are holomorphic when $w \in \Omega$. These kernels are called Henkin-Ramirez (H-R) kernel or Stein-Kerzman (S-K) kernel.

Alt [1] in 1974 and Kerzman and Stein [19] in 1978 proved respectively the Plemelj's formula for the H-R kernel and S-K kernel. They proved: if Ω is C^∞ smooth, bounded strictly pseudoconvex domain, $u \in C^\infty(\partial\Omega)$, $H(w,z)$ is H-R kernel or S-K kernel, and let

$$(1.4.1) \qquad Hu(w) = \int_{\partial\Omega} H(w,z)u(z)d\sigma_z, \qquad w \in \Omega,$$

then when w approaches $w_0 \in \partial\Omega$ along a non-tangential direction,

$$\lim_{w \to w_0} Hu(w) = \frac{1}{2}u(w_0) + \mathrm{P.\,V.}\int_{\partial\Omega} H(w_0,z)u(z)d\sigma_z$$

holds, where the principal value is defined as:

$$\mathrm{P.\,V.}\int_{\partial\Omega} H(w_0,z)u(z)d\sigma_z = \lim_{\varepsilon \to 0}\int_{\partial\Omega - B(w_0,\varepsilon)} H(w_0,z)u(z)d\sigma_z$$

and $B(w_0,\varepsilon) = \{z \in \partial\Omega \mid |g(z,w_0)| < \varepsilon\}$.

In 1982, Gong and Shi proved the following general Plemelj's formula.

THEOREM 1.4.1 [12]. *Suppose Ω is a smooth, bounded strictly pseudoconvex domain, and $H(w,z)$ is the H-R kernel or S-K kernel, $u \in C^\infty(\partial\Omega)$. Then*

$$H(w) = \int_{\partial\Omega} H(w,z)u(z)d\sigma_z, \qquad w \in \Omega$$

is holomorphic in Ω and admits a continuous extension up to $\bar{\Omega}$.

For $w_0 \in \partial\Omega$, define

$$(1.4.2) \qquad \mathrm{P.\,V.}\int_{\partial\Omega} H(w_0,z)u(z)d\sigma_z = \lim_{\varepsilon \to 0}\int_{b - D(w_0,\varepsilon)\cap\partial\Omega} H(w_0,z)u(z)d\sigma_z,$$

where $D(w_0,\varepsilon)$ is the neighborhood around w_0 and contracts to the point w_0 as $\varepsilon \to 0$. If

$$\lim_{\varepsilon \to 0}\lim_{\delta \to 0}\int_{D(w_0,\varepsilon)\cap\partial\Omega} H(w_0 + \delta\gamma, z)d\sigma_z = a$$

exists, where a is a constant and γ is the inner normal to $\partial\Omega$ at w_0. Then when w approaches $w_0 \in \partial\Omega$ along a non-tangential direction we have

$$\lim_{w \to w_0} Hu(w) = au(w_0) + H_a u(w_0),$$

where $H_a u(w_0)$ denotes the value of (1.4.2).

Obviously, this deduces to the Theorem of Alt [1] and Kerzman-Stein [19] when $D(w_0, \varepsilon)$ is $\{z \in \partial\Omega \mid |g(z, w_0)| < \varepsilon\}$. In this case, $a = \frac{1}{2}$ and Plemelj's formula becomes

$$\lim_{w \to w_0} Hu(w) = \frac{1}{2}u(w_0) + H_{\frac{1}{2}}u(w_0).$$

By taking the neighborhood to be an "ellipse" or a "rectangle" respectively:

$$D_e(w_0, \varepsilon) = \{z \in \partial\Omega \mid a^2(\operatorname{Re} g)^2 + \beta^2(\operatorname{Im} g)^2 \leqslant \varepsilon\},$$

$\alpha \geqslant 0$, $\beta \geqslant 0$, $\alpha + \beta \neq 0$;

$$D_R(w_0, \varepsilon) = \{z \in \Omega \mid |\operatorname{Re} g| < \alpha\varepsilon, |\operatorname{Im} g| \leqslant \beta\varepsilon\},$$

$\alpha > 0$, $\beta > 0$, they proved [19]:

(1.4.3)

$$\lim_{\varepsilon \to 0} \lim_{\delta \to 0} \int_{D_e(w_0, \varepsilon)} H(w_0 + \delta\gamma, z) d\sigma_z = \frac{1}{2}\left(\frac{2\beta}{\alpha + \beta}\right)^{n-1},$$

(1.4.4)

$$\lim_{\varepsilon \to 0} \lim_{\delta \to 0} \int_{D_R(w_0, \varepsilon)} H(w_0 + \delta\gamma, z) d\sigma_z = \frac{2^{n-1}}{\pi}\left(\frac{\pi}{2} - b\right),$$

where

$$b = \int_0^{\arctan \frac{\beta}{\alpha}} \cos^{n-2} t \frac{\sin(n-1)t}{\sin t} dt.$$

By (1.4.3) and (1.4.4), they obtained the corresponding Plemelj's formula [19].

§2. Function spaces

2.1 On coefficient inequalities

Let Ω be a bounded symmetric domain in the complex vector space $\mathbb{C}^n$, $0 \in \Omega$, with Bergman-Silov boundary b. By $H(\Omega)$ we denote the class of all holomorphic functions on Ω.

If $f \in H(\Omega)$, the means $M_p(r, f)$, $0 < p \leqslant \infty$, are defined by

$$M_p(r, f) = \left\{\int_b |f(r\zeta))|^p d\sigma(\zeta)\right\}^{\frac{1}{p}}, \qquad 0 < p < \infty,$$

$$M_\infty(r, f) = \sup\{|f(r\zeta)| \mid \zeta \in b\},$$

where σ is the Lebesgue measure on b with $\sigma(b) = 1$.

As usual, for $0 < p < \infty$, $H^p(\Omega)$ denotes the space of holomorphic functions on Ω for which the means $M_p(r, f)$ are bounded and the norm of $f \in H^p(\Omega)$ is defined by

$$\|f\|_p = \sup\{M_p(r, f) \mid 0 < r < 1\}.$$

Hua [17] constructed by group representation theory a system $\{\phi_{kv}\}$ of homogeneous polynomials, $k = 0, 1, \ldots$; $v = 1, \ldots, m_k$, $m_k = \dfrac{(n + k - 1)!}{k!(n - 1)!}$, complete and orthogonal on Ω and orthogonal on b. Every $f \in H(\Omega)$ has a series expansion [14]

$$(2.1.1) \qquad f(z) = \sum_{k,v} a_{kv}\phi_{kv}(z), \qquad a_{kv} = \lim_{r \to 1} \int_b f(r\zeta)\overline{\phi_{kv}(\zeta)} d\sigma(\zeta)$$

where $\sum_{k,v} = \sum_{k=0}^{\infty} \sum_{v=1}^{m_k}$ and the convergence is uniform on compact subsets of Ω.

In one variable, Hardy and Littlewood proved the following theorems.

THEOREM A [4]. *If $f(z) = \sum_{k=0}^{\infty} a_k z^k \in H^p$, $0 < p \leqslant 2$, then $\sum_{k=0}^{\infty}(k + 1)^{p-2}|a_k|^p < \infty$, and*

$$\left\{\sum_{k=0}^{\infty}(k + 1)^{p-2}|a_k|^p\right\}^{\frac{1}{p}} \leqslant C_p\|f\|_p,$$

where C_p depends only on p.

THEOREM B [4]. *Let $\{a_k\}$ be a sequence of complex numbers such that for some q, $2 \leqslant q < \infty$, $\sum_{k=0}^{\infty}(k + 1)^{q-2}|a_k|^q < \infty$. Then $f(z) = \sum_{k=0}^{\infty} a_k z^k \in H^q$, and*

$$\|f\|_q \leqslant C_q\left\{\sum_{k=0}^{\infty}(k + 1)^{q-2}|a_k|^q\right\}^{\frac{1}{q}},$$

where C_q depends only on q.

We generalized these two theorems to bounded symmetric domains of $\mathbb{C}^n$. We have

THEOREM 2.1.1 [35]. *Let Ω be a bounded symmetric domain. If $f(z) = \sum_{k,v} a_{kv}\phi_{kv}(z) \in H^p(\Omega)$, $0 < p \leqslant 2$, then $\sum_{k=0}^{\infty}(k + 1)^{n(p-2)}\sum_{v=1}^{m_k}|a_{kv}|^p < \infty$, and*

$$\sum_{k=0}^{\infty}(k + 1)^{n(p-2)}\sum_{v=1}^{m_k}|a_{kv}|^p \leqslant C_p\|f\|_p,$$

where C_p depends only on p.

THEOREM 2.1.2 [35]. *Let $\{a_{kv}\}$, $v = 1, \ldots, m_k$, $k = 0, 1, \ldots$, be a sequence of complex numbers such that*

$$\sum_{k=0}^{\infty}(k + 1)^{n(q-2)}\sum_{v=1}^{m_k}|a_{kv}|^q < \infty$$

for some q, $2 \leqslant q < \infty$. *Then* $f(z) = \sum_{k,v} a_{kv}\phi_{kv}(z) \in H^q(\Omega)$, *and*

$$\|f\|_q \leqslant C_q \left\{ \sum_{k=0}^{\infty} (k+1)^{n(q-2)} \sum_{v=1}^{m_k} |a_{kv}|^q \right\}^{\frac{1}{q}},$$

where C_q depends only on q.

More recently, we introduced the space $D^p(\Omega)$ as follows [36]: Let $f \in H(\Omega)$ with the expansion (2.1.1), denote

$$b_\ell^2 = \sum_{v=1}^{m_\ell} |a_{lv}|^2, \qquad I_k = \{s \mid 2^k \leqslant s < 2^{k+1}, s \in \mathbb{N}\},$$

we say that $f \in D^p(\Omega)$, $0 < p < \infty$, if

$$\|f\|_{D^p} = \sum_{k=0}^{\infty} 2^{-kn} \left(2^{kn} \sum_{\ell \in I_k} b_\ell^2 \right)^{\frac{p}{2}} < \infty.$$

The space $D^p(\Omega)$ is closely related with the Hardy $H^p(\Omega)$. We proved

THEOREM 2.1.3 [36]. (a) *If $0 < p < 2$, then $H^p(\Omega) \subset D^p(\Omega)$.*
(b) *If $p \geqslant 2$, then $D^p(\Omega) \subset H^p(\Omega)$.*

For the unit ball B of $\mathbb{C}^n$, we have

THEOREM 2.1.4 [36]. (a) *If $0 < p < 2$, then there exist $f \in D^p(B)$, but $f \notin H^p(B)$.*
(b) *If $p > 2$, then there exist $f \in H^p(B)$, but $f \notin D^p(B)$.*

It is an interesting fact that Theorems A and B can be further extended to $D^p(\Omega)$:

THEOREM 2.1.5 [36]. *If $f(z) = \sum_{k,v} a_{kv}\phi_{kv}(z) \in D^p(\Omega)$, $0 < p \leqslant 2$, then*

$$\sum_{k=0}^{\infty} (k+1)^{n(p-2)} \sum_{v=1}^{m_k} |a_{kv}|^p \leqslant C_p \|f\|_{D^p},$$

where C_p depends only on p.

THEOREM 2.1.6 [36]. *If $\{a_{kv}\}$ satisfies the condition of theorem 2.1.2, then $f(z) = \sum_{k,v} a_{kv}\phi_{kv}(z) \in D^q(\Omega)$, and*

$$\|f\|_{D^q} \leqslant C_q \left\{ \sum_{k=0}^{\infty} (k+1)^{n(q-2)} \sum_{v=1}^{m_k} |a_{kv}|^q \right\},$$

where C_q depends only on q.

2.2 Inequalities for the integral means of
subharmonic functions and their derivatives

Let $f \in H(\Omega)$ with the expansion (2.1.1). In [46], Stoll defined the following multiplier transformations $J^\beta f$ and $J_\beta f$ for $\beta > 0$ respectively by

$$(J^\beta f)(z) = \sum_{k,v} \frac{\Gamma(k+n+\beta)}{\Gamma(k+n)} a_{kv}\phi_{kv}(z)$$

and

$$(J_\beta f)(z) = \sum_{k,v} \frac{\Gamma(k+n)}{\Gamma(k+n+\beta)} a_{kv}\phi_{kv}(z),$$

and proved the following theorems.

THEOREM C [46]. *Let f be holomorphic in the unit ball B, α, β positive real numbers and $1 \leqslant s \leqslant \infty$.*

(a) *If $M_s(r,f) = O((1-r)^{-\alpha})$, then $M_s(r, J^\beta f) = O((1-r)^{-\alpha-\beta})$.*
(b) *If $M_s(r,f) = O((1-r)^{\alpha-\beta})$, then $M_s(r, J_\beta f) = O((1-r)^{-\alpha})$.*

THEOREM D [46]. *Let f be holomorphic in B, $1 \leqslant s \leqslant \infty$, $-1 < b < \infty$ and $\beta > 0$.*

(a) *For all a, $0 < a < \infty$,*

$$\int_0^1 (1-r)^{a\beta+b} M_s^a(r, J^\beta f)dr \leqslant C \int_0^1 (1-r)^b M_s^a(r, f)dr.$$

(b) *Conversely, for all a, $1 \leqslant a < \infty$,*

(2.2.1) $$\int_0^1 (1-r)^b M_s^a(r, J_\beta f)dr \leqslant C \int_0^1 (1-r)^{a\beta+b} M_s^a(r, f)dr.$$

For both theorems C and D, the problem which is still unsolved is the case $0 < s < 1$. In addition, Stoll conjectured that (2.2.1) is also true for the case $0 < a < 1$.

In [37], we proved that Theorems C and D hold on bounded symmetric domains for all s, $0 < s \leqslant \infty$, and (2.2.1) is also true for all a, $0 < a < \infty$. We first proved that these theorems hold for $f^{[\beta]}$ and $f_{[\beta]}$, which were defined as follows:

$$f^{[\beta]}(z) = \sum_{k,v} \frac{\Gamma(k+\beta+1)}{\Gamma(k+1)} a_{kv}\phi_{kv}(z),$$

$$f_{[\beta]}(z) = \sum_{k,v} \frac{\Gamma(k+1)}{\Gamma(k+\beta+1)} a_{kv}\phi_{kv}(z).$$

By the relations

$$J^\beta f = (f_{[n-1]})^{[n+\beta+1]} \quad \text{and} \quad J_\beta f = (f_{[n+\beta-1]})^{[n-1]},$$

we obtained the corresponding theorems for $J^\beta f$ and $J_\beta f$.

THEOREM 2.2.1 [**37**]. *Let* $f \in H(\Omega)$, $0 < r < 1$, $\alpha > 0$, $\beta > 0$ *and* $0 < s \leqslant \infty$.

(a) *If* $M_s(r, f) = O((1 - r)^{-\alpha})$, *then* $M_s(r, f^{[\beta]}) = O((1 - r)^{-\alpha-\beta})$.
(b) *If* $M_s(r, f) = O((1 - r)^{-\alpha-\beta})$, *then* $M_s(r, f_{[\beta]}) = O((1 - r)^{-\alpha})$.

THEOREM 2.2.2 [**37**]. *Let* $f \in H(\Omega)$, $0 < r < 1$, $\alpha > 0$, $\beta > 0$, *and* $0 < s \leqslant \infty$.

(a) *If* $M_s(r, f) = O((1 - r)^{-\alpha})$, *then* $M_s(r, J^\beta f) = O((1 - r)^{-\alpha-\beta})$.
(b) *If* $M_s(r, f) = O((1 - r)^{-\alpha-\beta})$, *then* $M_s(r, J_\beta f) = O((1 - r)^{-\alpha})$.

This generalizes Theorem C.

THEOREM 2.2.3 [**37**]. *Let* $f \in H(\Omega)$, $0 < s \leqslant \infty$, $-1 < b \leqslant \infty$ *and* $0 < a < \infty$. *Then for all* $\beta > 0$,

$$\int_0^1 (1 - r)^{a\beta+b} M_s^a(r, f^{[\beta]})dr \leqslant C \int_0^1 (1 - r)^b M_s^a(r, f)dr.$$
$$\int_0^1 (1 - r)^b M_s^a(r, f_{[\beta]})dr \leqslant C \int_0^1 (1 - r)^{a\beta+b} M_s^a(r, f)dr.$$

THEOREM 2.2.4 [**37**]. *Let* $f \in H(\Omega)$, $0 < s \leqslant \infty$, $-1 < b < \infty$, $0 < a < \infty$. *Then for all* $\beta > 0$,

$$\int_0^1 (1 - r)^{a\beta+b} M_s^a(r, J^\beta f)dr \leqslant C \int_0^1 (1 - r)^b M_s^a(r, f)dr.$$
$$\int_0^1 (1 - r)^b M_s^a(r, J_\beta f)dr \leqslant C \int_0^1 (1 - r)^{a\beta+b} M_s^a(r, f)dr.$$

This generalizes Theorem D.

Let $\alpha = (\alpha_1, \ldots, \alpha_n)$ be a multi-index, define

$$(D^\alpha f)(z) = \frac{\partial^{|\alpha|} f(z)}{\partial z_1^{\alpha_1} \ldots \partial z_n^{\alpha_n}}.$$

It is natural to ask whether Theorem 2.2.3 holds for $D^\alpha f$. More recently, we proved the following two theorems for the unit ball of $\mathbb{C}^n$.

THEOREM 2.2.5 [**38**]. *Let* $f \in H(\Omega)$, $0 < p < \infty$, $-1 < b < \infty$ *and* $0 < a < \infty$. *Then for any* $\alpha = (\alpha_1, \ldots, \alpha_n)$, *there exists a positive constant* K *independent of* f, *such that*

$$\int_0^1 (1 - r)^{a|\alpha|+b} M_p^a(r, D^\alpha f)dr \leqslant K \int_0^1 (1 - r)^b M_p^a(r, f)dr.$$

THEOREM 2.2.6 [**38**]. *Let* $f \in H(\Omega)$, $0 < p < \infty$, $-1 < b < \infty$ *and* $0 < a < \infty$. *Then for any positive integer* m, *we have*

$$\int_0^1 (1-r)^b M_p^a(r,f)dr \leqslant K\left\{ \sum_{|\alpha|<m-1} |(D^\alpha f)(0)|^a \right.$$
$$\left. + \sum_{|\alpha|=m} \int_0^1 (1-r)^{am+b} M_p^a(r,D^\alpha f)dr \right\},$$

where K *is independent of* f.

Let v be the Lebesgue measure on $\mathbb{C}^n$ with $v(B) = 1$ and define

$$\|f\|_p = \left\{ \int_B |f(z)|^p dv(z) \right\}^{\frac{1}{p}}$$

and

$$\|f\|_{m,p} = \sum_{|\alpha|\leqslant m-1} |(D^\alpha f)(0)| + \sum_{|\alpha|=m} \|T_a f\|_p,$$

where $(T_p f)(z) = (1 - |z|^2)^{|\alpha|}(D^\alpha f)(z)$.

As an application of Theorems 2.2.5 and 2.2.6, we proved the following theorem.

THEOREM 2.2.7 [**38**]. *Let* m *be a positive integer and* $f \in H(B)$. *Then* $f \in L^p(dv)$, $0 < p < \infty$, *if and only if all the functions* $(1-|z|^2)^m(D^\alpha f)(z)$ *with* $|\alpha| = m$ *are in* $L^p(dv)$. *Moreover* $\| \cdot \|_p$ *and* $\| \cdot \|_{m,p}$ *are equivalent norms on* $L^p \cap H(B)$.

In [**49**], Zhu proved the above theorem for the case $p \geqslant 1$. His method cannot deal with the case $0 < p < 1$. Using theorems 2.2.5 and 2.2.6, we can prove that Theorem 2.2.7 holds for all $p \in (0,\infty)$.

Stoll proved the following two theorems in [**46**]:

THEOREM E. *Let* $f = u + iv$ *be holomorphic in* B. *If*

$$M_p(r,u) = O((1-r)^{-\alpha}), \qquad 1 < p < \infty, \qquad \alpha > 0$$

then

$$M_p(r,v) = O((1-r)^{-\alpha}).$$

THEOREM F. *Let* $f = u + iv$ *be holomorphic in* B *with* $f(0)$ *real. Then, for* $-1 < b < \infty$ *and* $1 \leqslant p \leqslant \infty$, $0 < a < \infty$,

$$\int_0^1 (1-r)^b M_p^a(r,v)dr \leqslant K \int_0^1 (1-r)^b M_p^a(r,u)dr.$$

We have generalized Theorems E and F to bounded symmetric domain of $\mathbb{C}^n$ in [**37**], but the problem which is still unresolved is the case $0 < p < 1$.

In [**38**], we first proved the following inequality:

THEOREM 2.2.8 [38]. *Let $f = u + iv$ be holomorphic in the unit ball of $\mathbb{C}^n$ and $0 < p < 1$, $s \geqslant 0$. Then*

$$|f(z)|^p \leqslant K \left\{ |v(0)|^p + \int_B |u(w)|^p \frac{(1 - |w|^2)^s}{|1 - \langle z, w \rangle|^{n+s+1}} dv(w) \right\}$$

for $z \in B$.

When $n = 1$, this theorem has been proved in [25]. We generalized it to the unit ball of $\mathbb{C}^n$.

From Theorem 2.2.8, we have

THEOREM 2.2.9 [38]. *Let $f = u + iv$ be holomorphic in B with $v(0) = 0$. If $0 < q < 1$, $p \geqslant q$ and $s \geqslant 0$, then*

$$M_p^q(r, f) \leqslant K \int_0^1 \frac{(1 - \rho)^s}{(1 - r\rho)^{s+1}} M_p^q(\rho, u) d\rho.$$

Thus, we obtained

THEOREM 2.2.10 [38]. *Let $f = u + iv \in H(B)$ and $0 < p < \infty$. If*

$$M_p(r, u) = O((1 - r)^{-\alpha}), \qquad \alpha > 0$$

then

$$M_p(r, v) = O((1 - r)^{-\alpha}).$$

THEOREM 2.2.11 [38]. *Let $f = u + iv \in H(B)$ with $v(0) = 0$ and $0 < p \leqslant \infty$, $a > 0$, $b > -1$. Then*

$$\int_0^1 (1 - r)^b M_p^a(r, v) dr \leqslant K \int_0^1 (1 - r)^b M)p^a(r, u) dr.$$

Until now, Theorems 2.2.5, 2.2.6, 2.2.10 and 2.2.11 were proved in the unit ball. One might conjecture that these theorems are also true on arbitrary bounded symmetric domains.

2.3 Bloch functions and $BMOA$ functions

In [47], Timoney gave the definition of Bloch function on bounded homogeneous domains and proved many equivalent conditions that a holomorphic function become a Bloch function.

1. In his paper, Timoney asked whether or not there are Bloch functions on B_n which are not in $H^p(B_n)$ (for $n \geqslant 2$). Ryll and Wojtaszczyk [29] affirmatively answered this question. They constructed a Bloch function on B_n, which is not in $H^p(B_n)$ for any $p > 0$. A general question is: is there a Bloch function on B_n, which is not of bounded characteristic? Using the Ryll-Wojtaszczyk polynomials [29], Shi proved the following:

THEOREM 2.3.1 [**39**]. *There is a Bloch function on B_n, which has a radial limit almost nowhere on ∂B_n.*

It is clear that the above function is not of bounded characteristic.

The following theorem plays an important role in the proof of Theorem 2.3.1, which also has some interest in its own right.

THEOREM 2.3.2 [**39**]. *Let p_m be the Ryll-Wojtaszczyk polynomials and ϕ_m the Rademacher functions. If $\sum_{m=1}^{\infty} a_m p_m(z)$ is holomorphic in B_n and $\sum_{m=1}^{\infty} |a_m|^2 = \infty$, then there exist unitary U_m such that for almost every $t \in [0,1]$,*

$$f_t(z) = \sum_{m=1}^{\infty} a_m \phi_m(t) p_m(zU_m)$$

has a radial limit almost nowhere on ∂B_n.

One might ask, under what conditions, a Bloch function on B_n belong to $H^p(B_n)$.

Hou obtained the following

THEOREM 2.3.3 [**16**]. *Let f be a Bloch function on B_n.*

(a) *If $0 < p < \infty$ and $\int_{B_n} |f(z)|^{p-2}|(\nabla f)(z)|dv(z) < \infty$, then $f \in H^p(B_n)$.*

(b) *If $2 < p < \infty$ and $\int_{B_n} |f(z)|^{p-2}|(Rf)(z)|dv(z) < \infty$, then $f \in H^p(B_n)$.*

On the boundary behavior of Bloch function, Hou proved the following interesting result.

THEOREM 2.3.4 [**16**]. *Let f be a Bloch function on B_n. If f has a radial limit almost everywhere on ∂B_n, then it has a K-limit almost everywhere on ∂B_n.*

As an application of Theorem 2.3.4, Hou obtained:

THEOREM 2.3.5 [**16**]. *Let $f(z) = \sum a_\alpha z^\alpha$ be a Bloch function on B_n. Assume it has a radial limit almost everywhere. Then*

$$\frac{1}{m^2} \sum_{|\alpha| \leqslant m} |\alpha|^2 |a_\alpha|^2 \omega_\alpha \to 0, \qquad (m \to \infty)$$

where $\omega_\alpha = \int_{\partial B_n} |\zeta^\alpha|^2 d\sigma(\zeta) = \dfrac{(n-1)!\alpha!}{(n-1+|\alpha|)!}.$

When $n = 1$, this result has been proved in [**26**].

2. In one variable, a holomorphic function $f : U \to \mathbb{C}$, U the unit disc, is a Bloch function if and only if $f(z) = c \log g'(z)$ for some univalent holomorphic function $g : U \to \mathbb{C}$ and some constant c. About this, Timoney said in [**47**]: "It seems hard to imagine a natural generalization of this to several variables."

We generalized this partly to several variables in the following way:

THEOREM 2.3.6 [40]. *Let U^n be the unit polydisc of $\mathbb{C}^n$. If $g : U^n \to \mathbb{C}^n$ is a convex biholomorphic map, $g(0) = 0$, then*

$$f(z) = \log \det g'(z)$$

is a Bloch function on U^n, where $g'(z)$ is the Jacobian matrix of g.

More recently, Liu proved the following:

THEOREM 2.3.7 [21]. *Let $g : B_n \to \mathbb{C}^n$ be a holomorphic map, $\phi \in \mathrm{Aut}(B_n)$ with $\phi(0) = 0$, and $F(w)$ be the normalization of $g(\phi(w))$, write*

$$F(w) = w + (wA^{(1)}w', \ldots, wA^{(n)}w') + \ldots,$$

where $w = (w_1, \ldots, w_n)$ and w' is the transpose of w.

(a) *If we denote $A^{(\ell)} = \left(a_{jk}^{(\ell)} \right)_{1 \leqslant j,k \leqslant n}$, then*

$$\frac{d}{d\rho} \log \det g'(\rho z) = \frac{(n+1)\rho|z|^2}{1 - \rho^2|z|^2} - \frac{2|z|}{1 - \rho^2|z|^2} \sum_{\ell,j=1}^{n} a_{\ell j}^{(\ell)} \frac{z_j}{|z|}.$$

(b) *If g is a convex map, then*

$$\left| \sum_{\ell,j=1}^{n} a_{\ell j}^{(\ell)} \frac{z_j}{|z|} \right| \leqslant C_n,$$

where C_n depends only on n.

From theorem 2.3.7, we have immediately

THEOREM 2.3.8. *If $g : B_n \to \mathbb{C}^n$ is a convex holomorphic map, then*

$$f(z) = \log \det g'(z)$$

is a Bloch function on B_n.

It is easy to see that if g fails to be convex, then Theorems 2.3.7 and 2.3.8 may be not true.

An interesting problem is whether every Bloch function f on B_n can be represented as $f(z) = c \log \det g(z)$ for some biholomorphic map g and some constant c.

3. It is well known that the main characterization of BMO functions, shown by John and Nirenberg [18], is that their distribution function possesses an exponential decay effect. In [2], for a bounded subset of the Nevanlinna class in the unit disk, Bearnstein proved that the distribution function of the non-tangential maximal function decreases in an exponential way. As a corollary, the John-Nirenberg theorem in the sense of the harmonic measure was obtained with a weaker integrability assumption. Ouyang [23, 24] generalized these results to the unit ball for a bounded subset of the H_ϕ class wider than the Nevanlinna class.

Let $\phi : [-\infty, \infty) \to (0, \infty)$ be a nondecreasing convex function, not identically 0. Define

$$H_\phi(B) = \left\{ f \,\middle|\, f \in H(B), I_1(f) = \sup_{0 < r < 1} \int_{\partial B} \phi(\log |f(r\zeta)|) d\sigma(\zeta) < \infty \right\}$$

and

$$M(f) = \{ g \mid g(z) = f \circ \phi(z) - f \circ \phi(0), \phi \in \mathrm{Aut}(B) \}.$$

Let E be a measurable set on ∂B, for $a \in B$, the harmonic measure of E at a with respect to B is defined by

$$\mu_a(E) = \int_B P(a, \zeta) d\sigma(\zeta),$$

where $P(a, \zeta)$ is the Poisson kernel of B.

THEOREM 2.3.9 [23, 24]. *Suppose $f \in H(B)$. If $M(f)$ is a bounded subset of H_ϕ, then*

$$\mu_0(\{ \eta \in \partial B \mid (M_\alpha g)(\eta) > t \}) < K e^{-\lambda t},$$

for any $g \in M(f)$, where K denotes an absolute constant, $\lambda = C \exp\{-\phi^{-1}(C(\alpha) \sigma_\phi(f))\}$, $\sigma_\phi(f) = \sup\{ I_1(g) \mid g \in M(f) \}$.

For the Nevanlinna class N, the converse is also true.

THEOREM 2.3.10 [23, 24]. *Let $f \in H(B)$. Then the following are equivalent:*

(a) *$M(f)$ is bounded in N.*
(b) *For every $g \in M(f)$, there exists an absolute constant K and a constant $\lambda = \lambda(\alpha, f)$, such that*

$$\mu_0(\{ \eta \in \partial B \mid (M_\alpha g)(\eta) > t \}) < K e^{-\lambda t}$$

for every $t > 0$.

From these, one can obtain the John-Nirenberg theorem in the sense of harmonic measure for $BMOA$ functions.

THEOREM 2.3.11 [23, 24]. *$f \in BMOA$ if and only if for every $a \in B$, there is*

$$\mu_a(\{ \zeta \in \partial B \mid |f(\zeta) - f(a)| > t \}) < K e^{-\lambda_1 t},$$

where K is an absolute constant.

For the relation between Bloch functions and $BMOA$ functions, Hou obtained the following.

THEOREM 2.3.12 [16]. *Let π be the orthogonal projection of $\mathbb{C}^n$ onto $\mathbb{C}^{n-1}$: $\pi(z_1,\ldots,z_n) = (z_1,\ldots,z_{n-1})$. If g is a Bloch function on B_{n-1}, then $g \circ \pi$ is a BMOA function on B_n.*

For $a \in B$, we define

$$B_1(a,r) = \{z \in \mathbb{C}^n \mid |z-a| < r, 0 < r \leqslant 1 - |a|\},$$
$$B_2(a,r) = \phi_a(B_1(0,r)),$$

where $\phi_a \in \mathrm{Aut}(B)$ with $\phi_a(a) = 0$. Denote

$$f_{B_j(a,r)} = \frac{1}{V(B_j(a,r))} \int_{B_j(a,r)} f(z)dv(z), \qquad j = 1,2,$$

and define

$$BMOA_j(B) = \left\{ f \in H(B) \, \Big| \, \|f\|_{BMOA} = \sup_{B_j \subset B} \frac{1}{V(B_j(a,r))} \right.$$
$$\left. \times \int_{B_j(a,r)} |f(z) - f_{B_j(a,r)}|dv(z) < \infty \right\},$$

$j = 1,2$. Hou proved

THEOREM 2.3.13 [16]. *$BMOA_1(B) = BMOA_2(B)$ and there exist $C_1, C_2 > 0$, such that*

$$C_1\| \cdot \|_{BMOA_1} \leqslant \| \cdot \|_{BMOA_2} \leqslant C_2\| \cdot \|_{BMOA_1}.$$

2.4 Representations of linear functionals

In [5], Duren, Romberg and Shields gave the representations of continuous linear functionals on $H^p(U)$, $0 < p < 1$, U being the unit disc. They also introduced the space B^p and proved that H^p is dense in B^p and both have the same dual spaces. Mitchell and Hahn [22] generalized the above results to the unit ball of $\mathbb{C}^n$. They also introduced the space $B_{p1}(B)$, $0 < p < 1$, and proved that $H^p(B)$ is dense in $B_{p1}(B)$, but they had not answered whether both $H^p(B)$ and $B_{p1}(B)$ have the same dual space. Using the methods in [30], we proved

THEOREM 2.4.1 [43]. *If $0 < p < q \leqslant \infty$, then $(B_{pq}(B))^* = (H^p(B))^*$.*

As an essential prerequisite for proving Theorem 2.4.1, we proved the following fact which has some independent interest.

THEOREM 2.4.2 [34]. *$H^p(B)$, $0 < p < 1$, and $B_{pq}(B)$, $0 < p < q < 1$, are not locally convex.*

Let Φ denote the class of functions $\phi : [0,1] \to [0,\infty)$ satisfying the following conditions:

(i) ϕ is continuous, increasing and $\phi(0) = 0$, $\phi(t) \neq 0$ if $t \neq 0$,
(ii) $\dfrac{t}{\phi(t)}$ is increasing and $\dfrac{t}{\phi(t)} \to 0$ as $t \to 0$,

(iii) $\int_0^\delta \frac{\phi(t)}{t}dt = O(\phi(\delta))$ and $\int_\delta^1 \frac{\phi(t)}{t^2}dt = O\left(\frac{\phi(\delta)}{\delta}\right)$ for $\delta > 0$.

Let $\phi \in \Phi$. A function $f \in A(B)$ is said to be of class $\Lambda_\phi(B)$ if its boundary function satisfies Lipschitz condition

$$|f(e^{i(\theta+h)}\zeta) - f(e^{i\theta}\zeta)| \leqslant K|\phi(h)|$$

for $\zeta \in \partial B$ and $\theta, h \in \mathbb{R}$. The norm of $f \in \Lambda_\phi(B)$ is defined by

$$\|f\|_{\Lambda_\phi} = \sup_{z \in B} \frac{1-|z|}{\phi(1-|z|)}|f^{[1]}(z)|.$$

We say that f is of class $\lambda_\phi(B)$ if $f \in \Lambda_\phi(B)$ and

$$\frac{1-|z|}{\phi(1-|z|)}|f^{[1]}(z)| \to 0 \qquad \text{as } |z| \to 1.$$

Let $\phi \in \Phi$. A function g holomorphic in B is said to be of class $\Gamma_\phi(B)$ if

$$\|g\|_{\Gamma_\phi} = \int_0^1 \phi(1-r)M_1(r, g^{[1]})dr < \infty.$$

Let $f(z) = \sum_{\alpha \geqslant 0} a_\alpha z^\alpha$ and $g(z) = \sum_{\alpha \geqslant 0} b_\alpha z^\alpha$, we denote

$$(f, g) = \lim_{r \to 1} \sum_{k=0}^\infty \left(\sum_{|\alpha|=k} a_\alpha \bar{b}_\alpha \omega_\alpha\right)r^k,$$

if the limit exists.

On the dual spaces $\Gamma_\phi^*(B)$ and $\lambda_\phi^*(B)$, we have

THEOREM 2.4.3 [44]. (i) *For every* $T \in \Gamma_\phi^*(B)$, *there exists a unique* $f \in \Lambda_\phi(B)$ *such that*

$$T(g) = (g, f)$$

for every $g \in \Gamma_\phi(B)$, *and*

$$C'\|f\|_{\Lambda_\phi} \leqslant \|T\|_{\Gamma_\phi^*} \leqslant C\|f\|_{\Lambda_\phi},$$

where the constants C *and* C' *are independent of* f *and* T.

(ii) *Conversely, for every* $f \in \Lambda_\phi(B)$,

$$T_f(g) = (g, f)$$

defines a bounded linear functional on $\Gamma_\phi^*(B)$, *and*

$$\|f\|_{\Gamma_\phi^*} \leqslant C\|f\|_{\Lambda_\phi},$$

where the constant C *is independent of* f.

THEOREM 2.4.4 [44]. (i) *For every* $F \in \lambda_\phi^*(B)$, *there exists a unique* $g \in \Gamma_\phi(B)$ *such that*

$$F(f) = (f, g)$$

for every $f \in \lambda_\phi(B)$, *and*

$$C'\|g\|_{\Gamma_\phi} \leqslant \|F\|_{\lambda_\phi^*} \leqslant C\|g\|_{\Gamma_\phi}.$$

(ii) *Conversely, for every* $g \in \Gamma_\phi(B)$,

$$F_g(f) = (f, g)$$

defines a bounded linear functional on $\lambda_\phi(B)$, *and*

$$\|F_g\|_{\lambda_\phi^*} \leqslant C\|g\|_{\Gamma_\phi}.$$

For $(N^+(B))^*$, Liu proved

THEOREM 2.4.5 [20]. (i) *For every* $T \in (N^+(B))^*$, *there exists a unique* $g(z) = \sum b_\alpha z^\alpha \in A(B)$, *such that*

$$(2.4.1) \qquad\qquad T(f) = \lim_{r \to 1} \int_{\partial B} f(r\zeta) g(\bar\zeta) d\sigma(\zeta)$$

for every $f \in N^+(B)$, *and*

$$b_\alpha = \frac{1}{\sqrt{\omega_\alpha}} O(\exp(-\delta|\alpha|^{\frac{n}{n+1}})),$$

where δ *is a positive constant.*

(ii) *Conversely, if* b_α *satisfies the above condition, then* $g(z) = \sum_{\alpha \geqslant 0} b_\alpha z^\alpha \in A(B)$, *and the equality* (2.4.1) *determines a bounded linear functional on* $N^+(B)$.

A positive continuous function ϕ on $[0, 1]$ is normal, if there exist $0 < a < b$ such that

(i) $\dfrac{\phi(r)}{(1-r)^a}$ is non-increasing and $\lim_{r \to 1} \dfrac{\phi(r)}{(1-r)^a} = 0$,

(ii) $\dfrac{\phi(r)}{(1-r)^b}$ is non-decreasing and $\lim_{r \to 1} \dfrac{\phi(r)}{(1-r)^b} = \infty$.

The functions $\{\phi, \psi\}$ will be called a normal pair if ϕ is normal and if for some b in (ii), there exists $\beta > b$, such that

$$\phi(r)\psi(r) = (1-r)^\beta, \qquad 0 \leqslant r < 1,$$

β is called the index of the normal pair $\{\phi, \psi\}$.

For $0 < p \leq \infty$, $0 < q \leqslant \infty$ and a normal function ϕ, let $H_{p,q}(\phi)$ denote the space of holomorphic functions f on B with

$$\|f\|_{p,q,\phi} = \left\{ \int_0^1 (1-r)^{-1}\phi^p(r) M_q^p(r, f) dr \right\}^{\frac{1}{p}} < \infty, \qquad 0 < p < \infty,$$

$$\|f\|_{\infty,q,\phi} = \sup_{0 \leqslant r < 1} \phi(r) M_q(r, f) < \infty, \qquad p = \infty.$$

Jevtic gave the dual spaces of $H_{p,q}(\phi)$ when $1 \leqslant p < \infty$ and $1 \leqslant q \leqslant \infty$. He obtained

THEOREM G. *Let* $1 \leqslant p < \infty$ *and* $1 \leqslant q \leqslant \infty$. *The dual of* $H_{p,q}(\phi)$ *can be identified with* $H_{p',q'}(\psi)$. *The pairing is given by the following*

$$(2.4.2) \qquad (f,g) = \int_B f(z)\overline{g(z)}(1-|z|^2)^{\beta-1}dv(z),$$

where β *is the index of the normal pair* $\{\phi, \psi\}$ *and* p', q' *are the conjugates of* p, q *respectively.*

Shi gave the dual of $H_{p,q}(\phi)$ for the other cases of p, q.

THEOREM 2.4.6 [45]. *If* $0 < p \leqslant 1$ *and* $1 < q \leqslant \infty$, *then*

$$(h_{p,q}(\phi))^* = H_{\infty,q'}(\psi).$$

The pairing is given by (2.4.2).

THEOREM 2.4.7 [45]. *If* $0 < p \leqslant 1$ *and* $0 < q \leqslant 1$, *then*

$$(H_{p,q}(\phi))^* = H_{\infty,\infty}(\psi).$$

The pairing is given by

$$(f,g) = \int_B f(z)\overline{g(z)}(1-|z|^2)^{\beta+\frac{n}{q}-n-1}dv(z),$$

where β *is the index of the normal pair* $\{\phi, \psi\}$.

2.5 Hadamard products and coefficient multipliers

In one variable it is well known that if

$$f(z) = \sum_{k=0}^{\infty} a_k z^k \qquad \text{and} \qquad g(z) = \sum_{k=0}^{\infty} b_k z^k$$

are two functions holomorphic in the unit disc, then the Hadamard product $f*g$ of f and g is defined as

$$(f*g)(z) = \sum_{k=0}^{\infty} a_k b_k z^k,$$

and many interesting properties of $f*g$ have been obtained.

In [42], Shi generalized this concept to functions holomorphic in the unit ball of $\mathbb{C}^n$.

Let

$$f(z) = \sum_{\alpha \geqslant 0} a_\alpha z^\alpha \qquad \text{and} \qquad g(z) = \sum_{\alpha \geqslant 0} b_\alpha z^\alpha$$

be two functions holomorphic in B, we define the Hadamard product $f*g$ of f and g as follows:

$$(f*g)(z) = \sum_{\alpha \geqslant 0} a_\alpha b_\alpha \omega_\alpha z^\alpha,$$

where
$$\omega_\alpha = \int_{\partial B} |\zeta^\alpha|^2 d\sigma(\zeta) = \frac{(n-1)!\alpha!}{(n+|\alpha|-1)!}.$$
From the definition, it is not hard to obtain the following integral representation of $f * g$:
$$(f * g)(z) = \int_{\partial B} f(r\zeta)g(r^{-1}z\bar{\zeta})d\sigma(\zeta), \qquad 0 < r < 1,$$
where $z\zeta = (z_1\zeta_1, \ldots, z_n\zeta_n)$. The above representation holds for every $z \in U^n$, and therefore $f * g$ is holomorphic in the unit polydisc U^n.

The following are the properties of $f * g$ on H^p spaces.

THEOREM 2.5.1 [42]. If $f \in H^p(B)$, $g \in H^q(B)$, and $p \geqslant 1$, $q \geqslant 1$, $\frac{1}{s} = \frac{1}{p} + \frac{1}{q} - 1 > 0$, then $f * g \in H^s(B)$, and

$$\|f * g\|_s \leqslant \|f\|_p \|g\|_q.$$

If p and s are required to satisfy $p \leqslant 2 \leqslant s$, the condition of Theorem 2.5.1 may be replaced by a weaker one.

THEOREM 2.5.2 [42]. Let $p \geqslant 1$, $q \geqslant 1$, $\frac{1}{s} = \frac{1}{p} + \frac{1}{q} - 1 > 0$ and $p \leqslant 2 \leqslant s$, $\beta \geqslant 0$. If $f \in H^p(B)$ and $M_q(r, R^\beta g) \leqslant K(1-r)^{-\beta}$, then $f * g \in H^s(B)$ and

$$\|f * g\|_s \leqslant K A(p, q, \beta)\|f\|_p.$$

Here $(R^\beta g)(z) = \sum_{\alpha \geqslant 0} |\alpha|^\beta b_\alpha z^\alpha$ if $g(z) = \sum_{\alpha \geqslant 0} b_\alpha z^\alpha$.

THEOREM 2.5.3 [42]. Let $p \geqslant 1$, $q \geqslant 1$, $\frac{1}{s} = \frac{1}{p} + \frac{1}{q} - 1 > 0$, $0 < \lambda < \beta$, and $\frac{n}{\ell} = \frac{n}{p} + \beta - \lambda$. If $f \in H^\ell(B)$ and $M_q(r, R^\beta g) \leqslant K(1-r)^{-\lambda}$, it follows that

(a) If $s < \infty$ then $f * g \in H^s(B)$ and $\|f * g\|_s \leqslant K A(p, q, \lambda)\|f\|_\ell$.
(b) If $s = \infty$ and $\ell \leqslant 1$ then $f * g \in C(\bar{B})$.

The following theorem gives a new characterization of Bloch functions on B. We denote the class of Bloch functions on Ω by $\beta(\Omega)$.

THEOREM 2.5.4 [42]. If $g \in H(B)$, then $g \in \beta(B)$ if and only if $f * g \in \beta(U^n)$ for every $f \in H^1(B)$.

Let X and Y be two spaces of holomorphic functions. A complex sequence $\{\lambda_\alpha\}$, where $\alpha = (\alpha_1, \ldots, \alpha_n)$ is a multi-index, is said to be a multiplier of X into Y, if $\sum a_\alpha z^\alpha \in X$ implies $\sum \lambda_\alpha a_\alpha z^\alpha \in Y$. By (X, Y) we denote the class of multipliers of X into Y.

It is an interesting problem to determine (X, Y) for some function spaces X and Y. In one variable, many authors [3] have obtained a lot of results about this topic. Here are some results on multipliers in several complex variables.

Using Theorem 2.5.1, we have

THEOREM 2.5.5 [42]. *Let $\{\lambda_\alpha\}$ be a complex sequence. If $\sum_{\alpha \geq 0} \lambda_\alpha \omega_\alpha^{-1} z^\alpha \in H^1(B)$, then $\{\lambda_\alpha\} \in (H^p(B), H^p(B))$ for every $p \geq 1$.*

Theorem 2.5.4 can be restated by means of the language of multiplier as follows:

THEOREM 2.5.4'. *$\{\lambda_\alpha\} \in (H^1(B), \beta(U^n))$ if and only if $\sum_{\alpha \geq 0} \lambda_\alpha \omega_\alpha^{-1} z^\alpha \in \beta(B)$.*

THEOREM 2.5.6 [41]. *Let $0 < p < 1 \leq q \leq \infty$ and $\beta > \frac{n}{p} - 1$. Then s sequence $\{\lambda_\alpha\} \in (H^p(B), H^q(B))$, if and only if for every $\zeta \in \partial B$, the function $g(z) = \sum_{\alpha \geq 0} \lambda_\alpha \omega_\alpha^{-1} z^\alpha$ satisfies*

$$M_q(r, g_\zeta^{[\beta]}) \leq K(1-r)^{n\left(\frac{1}{p}-1\right)-\beta},$$

where $g_\zeta(z) = g(z\zeta)$.

The space B_{pq} $(0 < p < q < \infty)$ is defined on B by

$$B_{pq}(B) = \{f \mid f \in H(B), \|f\|_{B_{pq}} < \infty\},$$

where

$$\|f\|_{B_{pq}} = \left\{ \int_0^1 (1-r)^{nq\left(\frac{1}{p}-\frac{1}{q}\right)-\beta} M_q^q(r, f) dr \right\}^{\frac{1}{q}}.$$

THEOREM 2.5.7 [41]. *Let $0 < p < 1$, $0 < q < 1$, and $\beta > \frac{n}{p} - 1$. Then $\{\lambda_\alpha\} \in (B_{p1}(B), B_{q1}(B))$ if and only if for every $\zeta \in \partial B$, $g(z) = \sum_{\alpha \geq 0} \lambda_\alpha \omega_\alpha^{-1} z^\alpha$ satisfies*

$$M_1(r, g_\zeta^{[\beta]}) \leq K(1-r)^{n\left(\frac{1}{p}-\frac{1}{q}\right)-\beta}.$$

Liu determined the spaces $(N^+(B), H^p)$ and $(N^+(B), A(B))$, where N^+ and $A(B)$ denote the Smirnov class and ball algebra respectively.

THEOREM 2.5.8 [20]. *$\{\lambda_\alpha\} \in (N^+(B), H^p(B))$, $0 < p \leq \infty$, or $\{\lambda_\alpha\} \in (N^+(B), A(B))$ if and only if*

$$\lambda_\alpha = O(\exp(-\delta|\alpha|^{\frac{n}{n+1}})),$$

where δ is a positive constant.

When $n = 1$, this result has been obtained by Yanagihara [48].

The problem which is unsolved is how to describe the space $(H^p(B), H^p(B))$, $0 < p < \infty$.

In order to give the space $(H_{p,q}(\phi), \ell^s)$, we need the following sequence spaces. Let $a = \{a_\alpha\}$ be a sequence, denote

$$\|a\|_{p,q} = \left\{ \sum_{m=0}^{\infty} \left(\sum_{k \in I_m} \sum_{|\alpha|=k} |a_\alpha|^p \right)^{\frac{q}{p}} \right\}^{\frac{1}{q}}, \qquad 1 \leqslant p < \infty, 1 \leqslant q < \infty.$$

$$\|a\|_{p,\infty} = \sup_m \left(\sum_{k \in I_m} \sum_{|\alpha|=k} |a_\alpha|^p \right)^{\frac{1}{p}}, \qquad 1 \leqslant p < \infty,$$

$$\|a\|_{\infty,q} = \left\{ \sum_{m=0}^{\infty} \left(\sup_{k \in I_m} \sum_{|\alpha|=k} |a_\alpha| \right)^q \right\}^{\frac{1}{q}}, \qquad 1 \leqslant q < \infty.$$

$$\|a\|_{\infty,\infty} = \sup_m \sup_{k \in I_m} \sum_{|\alpha|=k} |a_\alpha|,$$

where $I_m = \{k \mid k \text{ is an integer}, 2^{m-1} \leqslant k < 2^m\}$.

We define
$$\ell(p,q) = \{a = \{a_\alpha\} \mid \|a\|_{p,q} < \infty\}.$$

It is obvious that $\ell(p,p) = \ell^p$.

We proved the following

THEOREM 2.5.9 [45]. *Let* $p > 1$, $2 \leqslant q \leqslant \infty$ *and*

$$u = \begin{cases} \dfrac{2s}{2-s}, & s < 2, \\ \infty, & s \geqslant 2, \end{cases} \qquad v = \begin{cases} \dfrac{ps}{p-s}, & s < p, \\ \infty & s \geqslant p. \end{cases}$$

Then

$$(H_{p,q}(\phi), \ell^s) = \left\{ \{\lambda_\alpha\} \;\middle|\; \phi^{-1}\left(1 - \frac{1}{|\alpha|}\right) \lambda_\alpha \in \ell(u,v) \right\}.$$

THEOREM 2.5.10 [45]. *Let* $p > 1$. *Then*

$$(H_{p,1}(\phi), \ell^s) = \left\{ \{\lambda_\alpha\} \;\middle|\; \phi^{-1}\left(1 - \frac{1}{|\alpha|}\right) \lambda_\alpha \in \ell(s,t) \right\}.$$

where

$$t = \begin{cases} \dfrac{ps}{p-s}, & p > s, \\ \infty, & p \leqslant s. \end{cases}$$

REFERENCES

1. Alt, W., *Singulare Integrale mit gemischten Hologen eitaten auf Mannigfaltigkeiten und Anwendungen in der Funktionentheorie*, Math. Z. **137** (1974), 227–256.
2. Baerstein, A., *Aspects of contemporary complex analysis*, Academic Press, 1980, pp. 3–26.
3. Campbell, D. M. and Leach, R. J., *A survey of H^p multipliers as related to classical function theory*, Complex Variables **3** (1984), 85–111.
4. Duren, P. L., *Theory of H^p Spaces*, Academic Press, New York, 1970.
5. Duren, P. L., Romberg, B. W. and Shields, A. L., *Linear functionals on H^p spaces with $0 < p < 1$*, J. Reine Angew. Math. **238** (1969), 32–60.

6. Gong, S, *A remark on integrals of Cauchy type in several complex variables*, Proceedings of the First Symposia of Partial Differential Equation and Differential Geometry, Beijing, 1982.

7. Gong, S., *Singular Integrals in Several Complex Variables*, Shanghai Science and Technology Press, Shanghai, 1982.

8. Gong, S. and Sun J. G., *Integrals of Cauchy type in several complex varialbes*. I, Acta Mathematica Sinica **15** (1965), 431–443. (Chinese)

9. ———, *Integrals of Cauchy type in several complex variables*. II, Acta Mathematica Sinica **15** (1965), 775–799. (Chinese)

10. ———, *Integrals of Cauchy type in several complex variables*. III, Acta Mathematica Sinica **15** (1965), 800–811. (Chinese)

11. ———, *Singular integral equations on a cmplex hypersphere*, Acta Mathematica Sinica **16** (1966), 194–210. (Chinese)

12. Gong, S. and Shi, J. H., *Singular integrals in several complex variables*. I, Chin. Ann. of Math. **3** (1982), 483–502.

13. ———, *Singular integrals in several complex variables*. III, Chin. Ann. of Math. **4B** (1983), 467–484.

14. Hahn, K. T. and Mitchell, J., *H^p spaces on bounded symmetric domains*, Ann. Polon. Math. **28** (1973), 89–95.

15. Henkin, G., *Integral representations of functions holomorphic in strictly pseudoconvex domains and some applications*, Math. USSR, Sbornik **7** (1969), 597–616.

16. Hou, X. D., *Bloch functions on the ball*, Chinese Ann. of Math. **8A** (1987), 287–299.

17. Hua, L. K., *Harmonic Analysis of Functions of Several Complex Variables in the Classical Domains*, Transl. Math. Mono., vol. 6, Amer. Math. Soc., Providence, 1963.

18. John, F. and Nirenberg, L., *On functions of bounded mean oscillation*, Comm. on Pure and Appl. Math. **14** (1961), 415–426.

19. Kerzman, N. and Stein, E. M., *The Szego kernel in terms of Cauchy-Fantappie kernels*, Duke Math. J. **45** (1978), 197–224.

20. Liu, T. S., *Multipliers and linear functionals for the class N^+ in the unit ball* (to appear).

21. ———, *Distortion theorems on biholomorphic convex maps in the unit ball of $\mathbb{C}^n$* (to appear).

22. Mitchell, J. and Hahn, K. T., *Representation of linear functionals in H^p spaces over bounded symmetric domains in $\mathbb{C}^n$*, J. Math. Anal. Appl. **56** (1976), 379–396.

23. Ouyang, C. H., *Bearstein theorem in unit ball of $\mathbb{C}^n$*, Kexue Tongbao **32** (1987), 581–584.

24. ———, *Some classes of functions with exponential decay in the unit ball of $\mathbb{C}^n$*, Publ. RIMS, Kyoto Univ. **25** (1989), 263–277.

25. Pavlovic, M., *Mean values of harmonic conjugacies in the unit disc*, Complex Variables **10** (1988), 53–65.

26. Pommerenke, Ch., *On Bloch functions*, J. London Math. Soc. **2:2** (1970), 689–695.

27. Ramirez De Arellano, E., *Ein Division problem und Randinlegral-clarstellungen in der komplexen analysis*, Math. Ann. **184** (1970), 172–187.

28. Rudin, W., *Function Theory in the Unit Ball of $\mathbb{C}^n$*, Springer-Verlag, New York, 1980.

29. Ryll, J. and Wojtaszczyk, P., *The homogeneous polynomials on a complex ball*, Trans. Amer. Math. Soc. **276** (1983), 107–116.

30. Shapiro, J. H., *Mackey topologies, reproducing kernels and diagonal maps on the Hardy and Bergman spaces*, Duke Math. J. **43** (1976), 187–202.

31. Shi, J. H., *On the Cauchy type integrals for the hypersphere*, J. of University ofScience and Technology of China (USTC) **2** (1980), 1–9. (Chinese)

32. Shi, J. H. and Gong, S., *Singular integrals in several complex variables*. II, Chin. Ann. of Math. **4B:3** (1983), 307–318.

33. ———, *Singular integrals in several complex variables*. IV, Chin. Ann. of Math. **5B** (1984), 21–36.

34. Shi, J. H., *The spaces $H^p(B_n)$, $0 < p < 1$, and $B_{pq}(B_n)$, $0 < p < q < 1$, are not locally convex*, Proc. Amer. Math. Soc. **103** (1988), 69–74.

35. ———, *Hardy-Littlewood theorems on bounded symmetric domain of $\mathbb{C}^n$*, Scientia Sinica Ser. A **31** (1988), 916–926.

36. _____, D^p spaces on bounded symmetric domains of $\mathbb{C}^n$, Acta Math. Sinica (to appear).

37. _____, On the rate of growth of the means M_p of holomorphic and pluriharmonic functions on bounded symmetric domains of $\mathbb{C}^n$., J. Math. Anal. Appl. 126 (1987), 161–175.

38. _____, Inequalities for the integral means of holomorphic functions and their derivatives in the unit ball of $\mathbb{C}^n$, Trans. Amer. Math. Soc. 328 (1991), 619–637.

39. _____, Some applications of the Ryll-Wojtaszczyk polynomials, Acta Math. Sinica 1 (1985), 359–365.

40. _____, Two results on Bloch functions in several complex variables, J. of USTC 16 (1986), 147–152.

41. _____, Coefficient multipliers of H^p and B_{pq} spaces in the unit ball of $\mathbb{C}^n$ (to appear).

42. _____, Hadamard products of functions holomorphic in the unit ball of $\mathbb{C}^n$, Chinese J. of Contemporary Math. 9 (1988), 23–38.

43. _____, Mackey topologies of Hardy spaces and Bergman spaces in several complex variables, Chinese Science Bulletin 34 (1989), 1585–1589.

44. _____, Representation of linear functionals on Lipschitz spaces Λ_ϕ of functions holomorphic in the unit ball of $\mathbb{C}^n$, Chin. Ann. of Math. 8B (1987), 189–198.

45. _____, Duality and multipliers for mixed norm spaces in the ball (to appear).

46. Stoll, M., On the rate of growth of the means M_p of holomorphic and pluriharmonic functions in the ball, J. Math. Anal. Appl. 93 (1983), 109–127.

47. Timoney, R. M., Bloch functions in several complex variables. I, Bull. London Math. Soc. 12 (1980), 241–267.

48. Yanagihara, N., Multipliers and linear functionals for the class N^+, Trans. Amer. Math. Soc. 180 (1973), 449–461.

49. Zhu, K. H., The Bergman spaces, the Bloch spaces, and Cleason's problem, Trans. Amer. Math. Soc. 309 (1988), 253–268.

DEPARTMENT OF MATHEMATICS, UNIVERSITY OF SCIENCE AND TECHNOLOGY OF CHINA, HEFEI, ANHUI 230026, PEOPLE'S REPUBLIC OF CHINA

Contemporary Mathematics
Volume **142**, 1993

Some Results on the Homogeneous Siegel Domains in $\mathbb{C}^n$

YICHAO XU

It is well known that Vinberg, Pjatetski-Shapiro and Gindikin have proved that if D is a homogeneous bounded domain, then D is holomorphically isomorphic to a homogeneous Siegel domain which was defined by Pjatetski-Shapiro. In order to classify homogeneous bounded domains by holomorphic isomorphism, we have to give the realization of homogeneous Siegel domains. From this point of view, we will introduce the N-Siegel domain in section 1 and then summarize some of its properties obtained by the author in subsequent sections.

1. The Definition of N-Siegel Domains

Given an $N \times N$ symmetric matrix $S = (n_{ij})$ of non-negative integers such that

(1) $n_{ij} = n_{ji} > 0$, $n_{ii} = 1$, $\forall i, j$;
(2) If $n_{ik} = 0$, then $n_{ij}n_{jk} = 0$, $\forall j \in \{i+1, \ldots, k-1\}$.

A permutation σ of $1, 2, \ldots, N$ is said to be an admissible permutation of S, if $i < j$ and $\sigma(i) > \sigma(j)$ implies $n_{\sigma(i),\sigma(j)} = 0$. In this case, we denote $S^\sigma = (n_{\sigma(i)\sigma(j)})$.

Let A_{ij}^{tk}, $t = 1, 2, \ldots, n_{ij}$ be all $n_{ik} \times n_{jk}$ real matrices. The matrix set

$$\left\{ A_{ij}^{tk} \mid i \leqslant t \leqslant n_{ij}, 1 \leqslant i < j < k \leqslant N \right\}$$

is said to be an N-matrix set, if its matrices satisfy the following three conditions:

(3) $(A_{ij}^{sk})' A_{ij}^{tk} + (A_{ij}^{tk})' A_{ij}^{sk} = 2\delta_{st} I^{(n_{jk})}$, $1 \leqslant s, t \leqslant n_{ij}$, $1 \leqslant i < j < k \leqslant N$;
(4) $A_{ij}^{sk} A_{j\ell}^{tk} = \sum_r (e_r A_{ij}^{s\ell} e_t') A_{i\ell}^{rk}$, $1 \leqslant s \leqslant n_{ij}$, $1 \leqslant t \leqslant n_{j\ell}$;

1991 *Mathematics Subject Classification.* 32M10, 32H10, 32-02.

Research supported by the National Science Foundation of China.

This paper is in final form and no version of it will be submitted for publication elsewhere.

and

(5) $(A_{ij}^{sk})'A_{i\ell}^{tk} = \sum_r(e_t A_{ij}^{s\ell}e_r')A_{j\ell}^{rk}$, $1 \leqslant s \leqslant n_{ij}$, $1 \leqslant t \leqslant n_{i\ell}$, $1 \leqslant i < j < \ell < k \leqslant N$.

In the following we shall use m_i $(i = 1, 2, \ldots, N)$ to denote a set of N non-negative integers such that

(6) If $m_k = 0$, then $m_i n_{ki} = 0$, $\forall i \in \{k + 1, \ldots, N\}$.

Let $\{A_{ij}^{tk}, 1 \leqslant t \leqslant n_{ij}, 1 \leqslant i < j < k \leqslant N\}$ be an N-matrix set and Q_{ij}^t be all $m_i \times m_j$ complex matrices, $t = 1, 2, \ldots, n_{ij}$. The matrix set $\{A_{ij}^{tk}, Q_{ij}^t, 1 \leqslant t \leqslant n_{ij}, 1 \leqslant i < j < k \leqslant N\}$ is said to be an N-complex matrix set, if its matrices satisfy the following three conditions (7), (8) and (9):

(7) $(\overline{Q_{ij}^s})'Q_{ij}^t + (\overline{Q_{ij}^t})'Q_{ij}^s = 2\delta_{st}I^{(m_j)}$, $1 \leqslant s, t \leqslant n_{ij}$, $1 \leqslant i < j \leqslant N$;

(8) $Q_{ij}^s Q_{jk}^t = \sum_r(e_r A_{ij}^{sk}e_t')Q_{ik}^r$, $1 \leqslant s \leqslant n_{ij}$, $1 \leqslant t \leqslant n_{jk}$;

(9) $(\overline{Q_{ij}^s})'Q_{ik}^t = \sum_r(e_t A_{ij}^{sk}e_r')Q_{jk}^r$, $1 \leqslant s \leqslant n_{ij}$, $1 \leqslant t \leqslant n_{ik}$, $1 \leqslant i < j < k \leqslant N$.

Let $\{A_{ij}^{tk}, Q_{ij}^t\}$ and $\{B_{ij}^{tk}, T_{ij}^t\}$ be two N-complex matrix sets. And let S and $\tilde{S}$ be two symmetric matrices which are defined by $\{A_{ij}^{tk}\}$ and $\{B_{ij}^{tk}\}$ respectively. We say that $\{A_{ij}^{tk}, Q_{ij}^t\}$ and $\{B_{ij}^{tk}, T_{ij}^t\}$ are equivalent to each other if there exists an admissible permutation σ of S such that $\tilde{S} = S^\sigma$ and matrices $O_{ij} \in O(n_{ij})$, $1 \leqslant i < j \leqslant N$, $U_i \in U(m_i)$, $1 \leqslant i \leqslant N$ such that

$$B_{ij}^{tk} = \sum_r(e_r O_{\sigma(i)\sigma(j)}e_t')O'_{\sigma(i)\sigma(k)}A_{\sigma(i)\sigma(j)}^{r\sigma(k)}O_{\sigma(j)\sigma(k)},$$

$$T_{ij}^t = \sum_r(e_r O_{\sigma(i)\sigma(j)}e_t')\bar{U}'_{\sigma(i)}Q_{\sigma(i)\sigma(j)}^r U_{\sigma(j)},$$

$1 \leqslant t \leqslant n_{ij}$, $1 \leqslant i < j < k \leqslant N$.

DEFINITION. Given an N-matrix set $\{A_{ij}^{tk}\}$, let

$$n = N + \sum_{i<j} n_{ij},$$

and denote

$$x = (r_1, x_2, r_2, \ldots, x_N, r_N), \qquad x_j = (x_{1j}, x_{2j}, \ldots, x_{j-1,j}),$$

where $r_i \in \mathbb{R}$, $x_{ij} \in \mathbb{R}^{n_{ij}}$. Let

$$C_j(x) = \begin{pmatrix} r_i & x_{j,j+1} & \cdots & x_{jN} \\ x'_{j,j+1} & r_{j+1}I^{(n_{j,j+1})} & \cdots & R_{j+1,N}^j(x_{j+1,N}) \\ \vdots & \vdots & \ddots & \vdots \\ x'_{jN} & R_{j+1,N}^j(x_{j+1,N})' & \cdots & r_N I^{(n_{jN})} \end{pmatrix}$$

be an $n_j' \times n_j'$ matrix, $1 \leqslant j \leqslant N$, where $n_j' = \sum_{k=j}^N n_{jk}$ and

$$R_{k\ell}^j(x_{k\ell}) = \sum_r e_r' x_{k\ell}(A_{jk}^{r\ell})', \qquad 1 \leqslant j < k < \ell \leqslant N.$$

Then the point set in $\mathbb{R}^n$

$$\{x \in \mathbb{R}^n \mid C_1(x) > 0, \ldots, C_N(x) > 0\}$$

is said to be an N-cone.

LEMMA 1.1. *Let V_N be an N-cone defined by the N-matrix set $\{A_{ij}^{tk}\}$. Then V_N is an open convex cone which does not contain any straight lines. And V_N is a linear homogeneous cone.*

THEOREM 1.1. *Let V be an open convex cone which does not contain any straight lines. If V is linear homogeneous, then V is linearly isomorphic to an N-cone.*

THEOREM 1.2. *Let V_N be an N-cone and V_M an M-cone, defined by the N-matrix set $\{A_{ij}^{tk}\}$ and the M-matrix set $\{B_{ij}^{tk}\}$ respectively. Then V_N is linearly isomorphic to $\tilde{V}_M$ if and only if $M = N$ and the N-matrix sets $\{A_{ij}^{tk}\}$ and $\{B_{ij}^{tk}\}$ are equivalent to each other.*

In this case, V_N is linearly isomorphic to $\tilde{V}_N$ by

$$r_i \to r_{\sigma(i)}, \qquad x_{ij} \to x_{\sigma(i),\sigma(j)} O_{\sigma(i),\sigma(j)}.$$

Thus the classification of linear homogeneous cones can be reduced to the classification of N-matrix sets by the equivalent relation.

DEFINITION. Let V_N be an N-cone defined by the N-matrix set $\{A_{ij}^{tk}\}$. If

$$n_{1j} = n_{2j} = \cdots = n_{j-1,j} = \rho_j > 0, \qquad j = 2, 3, \ldots, N,$$

then V_N is said to be a square type cone. If

$$n_{i,i+1} = n_{i,i+2} = \cdots = n_{iN} = \sigma_i > 0, \qquad i = 1, 2, \ldots, N-1,$$

then V_N is said to be a dual square type cone.

The complete classification of square type cones and dual square type cones were obtained in 1978 and 1980. Note that the explicit basis of Clifford algebras and its total invariant were given in 1978.

DEFINITION. Given an N-complex matrix set $\{A_{ij}^{tk}, Q_{ij}^t\}$, let

$$n = N + \sum_{i<j} n_{ij}, \qquad m = \sum m_j,$$

and

$$z = (s_1, z_2, s_2, \ldots, z_N, s_N), \qquad z_j = (z_{ij}, \ldots, z_{j-1,j}), \qquad u = (u_1, \ldots, u_N),$$

where $s_i \in \mathbb{C}$, $z_{ij} \in \mathbb{C}^{n_{ij}}$, $u_i \in \mathbb{C}^{m_i}$. Let

$$P_j(u) = \begin{pmatrix} u_j \\ R^i_{j+1}(u) \\ \vdots \\ R^j_N(u) \end{pmatrix}.$$

Then the point set in $\mathbb{C}^{n+m}$ defined by

$$D(V_N, F) = \{(z, u) \in \mathbb{C}^n \times \mathbb{C}^m \mid$$
$$\mathrm{Im}(C_j(z)) - \mathrm{Re}(P_j(u)\overline{P_j(u)}') > 0, 1 \leqslant j \leqslant N\}$$

is said to be an N-Siegel domain.

LEMMA 1.2. *Let*

$$F(u, u) = (F_{11}(u, u), F_2(u, u), F_{22}(u, u), \ldots, F_N(u, u), F_{NN}(u, u)),$$
$$F_{jj}(u, u) = u_j \bar{u}'_j, \qquad F_j(u, u) = (F_{1j}(u, u), \ldots, F_{j-1,j}(u, u)),$$
$$F_{ij}(u, u) = (\mathrm{Re}(u_i Q^1_{ij} \bar{u}'_j), \ldots, \mathrm{Re}(u_i Q^{n_{ij}}_{ij} \bar{u}'_j)) \in \mathbb{R}^{n_{ij}}.$$

Then the N-Siegel domain $D(V_N, F)$ can be expressed as

$$D(V_N, F) = \{(z, u) \in \mathbb{C}^n \times \mathbb{C}^m \mid C_j(\mathrm{Im}(z) - F(u, u)) > 0, 1 \leqslant j \leqslant N\}$$
$$= \{(z, u) \in \mathbb{C}^n \times \mathbb{C}^m \mid \det C_j(\mathrm{Im}(z) - F(u, u)) > 0, 1 \leqslant j \leqslant N\}.$$

Obviously we have

THEOREM 1.3. *Any homogeneous Siegel domain is linearly isomorphic to an N-Siegel domain.*

Hence any homogeneous bounded domain is holomorphically isomorphic to an N-Siegel domain.

THEOREM 1.4. *Let $D(V_N, F)$ and $D(\tilde{V}_M, \tilde{F})$ be N- and M-Siegel domains defined by N- and M-complex matrix sets $\{A^{tk}_{ij}, Q^t_{ij}\}$ and $\{B^{tk}_{ij}, T^t_{ij}\}$ respectively. Then $D(V_N, F)$ is holomorphically isomorphic to $D(\tilde{V}_M, \tilde{F})$ if and only if $M = N$, and the N-complex matrix sets $\{A^{tk}_{ij}, Q^t_{ij}\}$, $\{B^{tk}_{ij}, T^t_{ij}\}$ are equivalent.*

In this case, $D(V_N, F)$ is linearly isomorphic to $D(\tilde{V}_N, \tilde{F})$ by

$$s_i \to s_{\sigma(i)}, \qquad z_{ij} \to z_{\sigma(i),\sigma(j)} O_{\sigma(i),\sigma(j)}, \qquad u_i \to u_{\sigma(i)} U_{\sigma(i)}.$$

It is well-known that if homogeneous Siegel domains D and $\tilde{D}$ are holomorphically isomorphic, then D and $\tilde{D}$ are linearly isomorphic. This result now follows from Theorem 1.4.

2. The Bergman Kernel, Cauchy-Szego Kernel and Formal Poisson Kernel

THEOREM 2.1. *Let*

$$D(V_N, F) = \{(z, u) \in \mathbb{C}^n \times \mathbb{C}^m \mid C_j(\mathrm{Im}(z) - F(u, u)) > 0, 1 \leqslant j \leqslant N\}$$

be an N-Siegel domain. Then its Bergman kernel is

$$K(z, u; \bar{z}, \bar{u}) = c_0 \prod_{j=1}^{N} (\det C_j(\mathrm{Im}(z) - F(u, u))^{\mu_j},$$

where c_0 is a positive constant and

$$m_j + 1 + \sum_{i=1}^{N} n_{ij} = -\sum_{i=1}^{j} \mu_i n_{ij}, \qquad j = 1, 2, \ldots, N.$$

DEFINITION. Let

$$D(V_N, F) = \{(z, u) \in \mathbb{C}^n \times \mathbb{C}^m \mid C_j(\mathrm{Im}(z) - F(u, U)) > 0, 1 \leqslant j \leqslant N\}$$

be an N-Siegel domain. A part boundary of $D(V_N, F)$:

$$\mathfrak{S}(V_N, F) = \{(z, u) \in \mathbb{C}^n \times \mathbb{C}^m \mid \mathrm{Im}(z) = F(u, u)\} \cong \mathbb{R}^n \times \mathbb{C}^m$$

is said to be the Silov boundary of $D(V_N, F)$.

THEOREM 2.2. *Let $D(V_N, F)$ be an N-Siegel domain. Then the Cauchy-Szego kernel is*

$$S(z, u; \xi, \eta) = c_0' \prod_{j=1}^{N} \det C_j \left(\frac{1}{2\sqrt{-1}}(z - \bar{\xi}) - F(u, \eta) \right)^{\nu_j},$$

where $(z, u) \in D(V_N, F)$, $(\xi, \eta) \in \mathfrak{S}(D(V_N, F))$, c_0' is a positive constant and

$$m_j + \frac{1}{2} + \frac{1}{2} \sum_{i=1}^{N} n_{ij} = -\sum_{i=1}^{j} \nu_i n_{ij}, \qquad j = 1, 2, \ldots, N.$$

DEFINITION. Let $D(V_N, F)$ be an N-Siegel domain. The function

$$P(z, u; \xi, \eta) = |S(z, u; \xi, \eta)|^2 / S(z, u; \bar{z}, \bar{u})$$

where $(z, u) \in D(V_N, F)$, $(\xi, \eta) \in \mathfrak{S}(D(V_N, F))$ is said to be the formal Poisson kernel.

DEFINITION. Let $D(V_N, F)$ be an N-Siegel domain. If $D(V_N, F)$ is a symmetric Hermitian manifold, then $D(V_N, F)$ is said to be a symmetric N-Siegel domain.

THEOREM 2.3. *Let $D(V_N, F)$ be an N-Siegel domain. Then the formal Poisson kernel $P(z, u; \xi, \eta)$ is a Poisson kernel if and only if $D(V_N, F)$ is a symmetric N-Siegel domain.*

The Bergman kernel, Cauchy-Szego kernel and Poisson kernel are given by L. K. Hua for the four classes of classical domains. On the other hand, Koranyi proved that the symmetric bounded domains and symmetric Siegel domains have Poisson kernel by the method of Lie groups. In a particular example of Q. K. Lu, it is proved that for certain non-symmetric homogeneous Siegel domains, the formal Poisson kernel is not a Poisson kernel. Later on E. M. Stein posed a conjecture: The formal Poisson kernel is not a Poisson kernel for any non-symmetric homogeneous bounded domains. This conjecture turns out to be true by Theorem 2.3.

3. The Exceptional Classical Domains

The realization of the exceptional classical domains in $\mathbb{C}^n$ is an important problem in the theory of Hermitian symmetric spaces. It was solved in 1978–1979.

The fifth class of exceptional classical domain $R_V(16)$ is $E_{6(-14)}/SO(2) \times SO(10)$, which is defined as a 2-Siegel domain, with $n_{12} = 6$, $m_1 = m_2 = 4$ and

$$Q_1 = I^{(4)},$$

$$Q_2 = \sqrt{-1} \begin{pmatrix} I^{(2)} & 0 \\ 0 & -I^{(2)} \end{pmatrix},$$

$$Q_3 = \begin{pmatrix} 0 & I^{(2)} \\ -I^{(2)} & 0 \end{pmatrix},$$

$$Q_4 = \sqrt{-1} \begin{pmatrix} 0 & \begin{pmatrix} 1 & 0 \\ 0 & -1 \end{pmatrix} \\ \begin{pmatrix} 1 & 0 \\ 0 & -1 \end{pmatrix} & 0 \end{pmatrix},$$

$$Q_5 = \begin{pmatrix} 0 & \begin{pmatrix} 0 & 1 \\ -1 & 0 \end{pmatrix} \\ \begin{pmatrix} 0 & 1 \\ -1 & 0 \end{pmatrix} & 0 \end{pmatrix},$$

$$Q_6 = \sqrt{-1} \begin{pmatrix} 0 & \begin{pmatrix} 0 & 1 \\ 1 & 0 \end{pmatrix} \\ \begin{pmatrix} 0 & 1 \\ 1 & 0 \end{pmatrix} & 0 \end{pmatrix}.$$

In this case,

$$C_1(z) = \begin{pmatrix} s_1 & z_{12} \\ z'_{12} & s_2 I^{(6)} \end{pmatrix},$$

$$C_2(z) = s_2,$$

$$P_1(u) = \begin{pmatrix} u_1 \\ \sum_{r=1}^{6} e'_r u \bar{Q}'_r \end{pmatrix},$$

$$P_2(u) = u_2,$$

where $z = (s_1, z_{12}, s_2) \in \mathbb{C} \times \mathbb{C}^6 \times \mathbb{C}$, $u = (u_1, u_2)$, and

$$R_V(16) = \left\{ (z, u) \in \mathbb{C}^{16} \mid \text{Im}(C_1(z)) - \text{Re}(P_1(u)\overline{P_1(u)}') > 0 \right\}.$$

That is

$$R_V(16) = \left\{ z \in \mathbb{C}^8, u_1, u_2 \in \mathbb{C}^4 \,\middle|\, \text{Im}(s_1) - u_1\bar{u}_1' > 0, \right.$$

$$\left. (\text{Im}(s_1) - u_1\bar{u}_1')(\text{Im}(s_2) - u_2\bar{u}_2') - \sum_{j=1}^{6}(\text{Im}(z_j) - \text{Re}(u_1 Q_j \bar{u}_2'))^2 > 0 \right\},$$

where $z_{12} = (z_1, \ldots, z_6) \in \mathbb{C}^6$.

Using Theorem 2.1, and noticing that $\mu_1 = -12$, $\mu_2 = 60$, the Bergman kernel of $R_V(16)$ becomes

$$K(z, u; \bar{z}, \bar{u}) = c_0 (\det C_1(\text{Im}\, z - F(u, u)))^{-12} (\det C_2(\text{Im}\, z - F(u, u)))^{60},$$

where

$$F_{11}(u, u) = u_1\bar{u}_1',$$
$$F_{22}(u, u) = u_2\bar{u}_2',$$
$$F_{12}(u, u) = (\text{Re}(u_1 Q_1 \bar{u}_2'), \ldots, \text{Re}(u_1 Q_6 \bar{u}_2')).$$

Hence

$$K(z, u; \bar{z}, \bar{u}) = c_0 \left[(\text{Im}(s_1) - u_1\bar{u}_1')(\text{Im}(s_2) - u_2\bar{u}_2') \right.$$

$$\left. - \sum_{j=1}^{6}(\text{Im}(z_j) - \text{Re}(u_1 Q_j \bar{u}_2'))^2 \right]^{-12}.$$

Using Theorem 2.2 and the fact that $\nu_1 = -8$, $\nu_2 = 40$, we find the Cauchy-Szego kernel of $R_V(16)$ to be

$$S(z, u; \xi, \eta) = c_0' \left[\left(\frac{1}{2\sqrt{-1}}(s_1 - \xi_1) - u_1\eta_1' \right) \left(\frac{1}{2\sqrt{-1}}(s_2 - \xi_2) - u_2\eta_2' \right) \right.$$

$$\left. - \sum_{j=1}^{6} \left(\frac{1}{2\sqrt{-1}}(z_j - \xi_j) - \frac{1}{2}u_1 Q_j \eta_2' - \frac{1}{2}u_2 \bar{Q}_j' \eta_1' \right)^2 \right]^{-8},$$

where $\xi = (\xi_1, \zeta_1, \ldots, \zeta_6, \xi_2)$, $\eta = (\eta_1, \eta_2)$ and

$$\text{Im}(\xi_1) = \eta_1\bar{\eta}_1', \qquad \text{Im}(\xi_2) = \eta_2\bar{\eta}_2', \qquad \text{Im}(\zeta_j) = \text{Re}(\eta_1 Q_j \bar{\eta}_2'), \qquad 1 \leqslant j \leqslant 6.$$

By using Theorem 2.3, we can derive the Poisson kernel of $R_V(16)$ to be

$$P(z, u; \xi, \eta) = |S(z, u; \xi, \eta)|^2 / S(z, u; \bar{z}, \bar{u}),$$

where

$$S(z, u; \bar{z}, \bar{u}) = c_0' \left[(\mathrm{Im}(s_1) - u_1 \bar{u}_1')(\mathrm{Im}(s_2) - u_2 \bar{u}_2') \right.$$

$$\left. - \sum_{j=1}^{6} (\mathrm{Im}(z_j) - \mathrm{Re}(u_1 Q_j \bar{u}_2'))^2 \right]^{-8}.$$

The holomorphic automorphism group of $R_V(16)$ is

$$\mathrm{Aut}(R_V(16)) = \mathrm{Iso}_{(\sqrt{-1}v_0, 0)}(R_V(16)) \cdot T(R_V(16)),$$

where $T(R_V(16))$ is a closed subgroup of $\mathrm{Aut}(R_V(16))$, which acts simply transitively on $R_V(16)$. Let $\mathfrak{t}(R_V(16))$ be the Lie algebra of $T(R_V(16))$. Then $\mathfrak{t}(R_V(16))$ has a basis

$$\frac{\partial}{\partial s_1}, \frac{\partial}{\partial s_2}, \frac{\partial}{\partial z_1}, \ldots, \frac{\partial}{\partial z_6},$$

$$A_1 = 2s_1 \frac{\partial}{\partial s_1} + z_{12} \frac{\partial'}{\partial z_{12}} + u_1 \frac{\partial'}{\partial u_1},$$

$$A_2 = 2s_2 \frac{\partial}{\partial s_2} + z_{12} \frac{\partial'}{\partial z_{12}} + u_2 \frac{\partial'}{\partial u_2},$$

$$X_t = 2z_t \frac{\partial}{\partial s_1} + s_2 \frac{\partial}{\partial z_t} + u_2 \bar{Q}_t' \frac{\partial'}{\partial u_1}, \qquad 1 \leqslant t \leqslant 6,$$

$$\alpha \frac{\partial'}{\partial u} + 2\sqrt{-1} F(u, \alpha) \frac{\partial'}{\partial z}, \qquad \forall \alpha \in \mathbb{C}^8.$$

The holomorphic isomorphism σ is defined as

$$r_1 = 2\sqrt{-3} + 4\sqrt{3}(s_2 + \sqrt{-1})\Delta^{-1},$$

$$r_2 = 2\sqrt{-3} + 4\sqrt{3}(s_1 + \sqrt{-1})\Delta^{-1},$$

$$x_{12} = 4\sqrt{6}\Delta^{-1} z_{12},$$

$$v_1 = 4\sqrt{-3}\Delta^{-1} \left[(s_1 + \sqrt{-1})u_1 - \sum_t z_t u_2 \bar{Q}_t' \right],$$

$$v_2 = 4\sqrt{-3}\Delta^{-1} \left[(s_1 + \sqrt{-1})u_1 - \sum_t z_t u_1 Q_t \right],$$

where $\Delta = (s_1 + \sqrt{-1})(s_2 + \sqrt{-1}) - zz'$ and $\sigma(\sqrt{-1}v_0, 0) = 0$. Let $\mathfrak{iso}_0(\sigma(R_V(16))$ be the Lie algebra of $\mathrm{Iso}_0(\sigma(R_V(16)))$. Then $\mathfrak{iso}_0(\sigma(R_V(16))$ has a basis

$$\{B_1, B_2, L_{ij}, L_0, Z_1(\alpha), Z_2(\beta), Z_3(\delta)\},$$

where

$$B_1 = \sqrt{-1} \left(2r_1 \frac{\partial}{\partial r_1} + x_{12} \frac{\partial'}{\partial x_{12}} + v_1 \frac{\partial'}{\partial v_1} \right),$$

$$B_2 = \sqrt{-1} \left(2r_2 \frac{\partial}{\partial r_2} + x_{12} \frac{\partial'}{\partial x_{12}} + v_2 \frac{\partial'}{\partial v_2} \right),$$

$$L_{ij} = x_i \frac{\partial}{\partial x_j} - x_j \frac{\partial}{\partial x_i} - \frac{1}{2} v_1 Q_i Q_j \frac{\partial'}{\partial v_1} + \frac{1}{2} v_2 Q_j Q_i \frac{\partial'}{\partial v_2},$$

$$1 \leqslant i < j \leqslant 6,$$

$$L_0 = \sqrt{-1} \left(v_1 \frac{\partial'}{\partial v_1} + v_2 \frac{\partial'}{\partial v_2} \right),$$

$$Z_1(\alpha) = \sqrt{-1} v_1 \bar{\alpha}' \frac{\partial}{\partial r_1} - \sum (v_2 \bar{Q}'_t \bar{\alpha}') \frac{\partial}{\partial x_t}$$

$$- \sqrt{2} r_1 \alpha \frac{\partial'}{\partial v_1} + \sum x_t \alpha Q_t \frac{\partial'}{\partial v_2},$$

$$Z_2(\beta) = \sqrt{-1} v_2 \bar{\beta}' \frac{\partial}{\partial r_2} - \sum (v_1 Q_t \bar{\beta}') \frac{\partial}{\partial x_t}$$

$$+ \sum x_t \beta \bar{Q}'_t \frac{\partial'}{\partial v_1} - \sqrt{2} r_2 \beta \frac{\partial'}{\partial v_2},$$

$$Z_3(\delta) = \sqrt{2} \left(x_{12} \delta' \frac{\partial}{\partial r_1} - x_{12} \delta' \frac{\partial}{\partial r_2} - r_1 \bar{\delta} \frac{\partial'}{\partial x_{12}} + r_2 \delta \frac{\partial'}{\partial x_{12}} \right)$$

$$- \sum \delta_t v_2 \bar{Q}'_t \frac{\partial'}{\partial v_1} + \sum \bar{\delta}_t v_1 Q_t \frac{\partial'}{\partial v_2},$$

and $x_{12} = (x_1, \ldots, x_6)$, $\delta = (\delta_1, \ldots, \delta_6)$, $\alpha, \beta \in \mathbb{C}^4$, $\delta \in \mathbb{C}^6$.

The sixth class of the exceptional classical domain $R_{VI}(27)$ is $E_{7(-25)}/T_x E_6$, which is defined as a 3-Siegel domain of the first kind, with $n_{12} = n_{13} = n_{23} = 8$, $m = 0$, and

$$R_{VI}(27) = \left\{ z \in \mathbb{C}^{27} \mid \mathrm{Im}(C_1(z)) > 0 \right\}$$
$$= \left\{ z \in \mathbb{C}^{27} \mid r_1 > 0, r_1 r_2 - |y_{12}|^2 > 0, \Delta(\mathrm{Im}(z)) > 0 \right\}$$

in which

$$z = (s_1, z_{12}, s_2, z_{13}, z_{23}, s_3) \in \mathbb{C} \times \mathbb{C}^8 \times \mathbb{C} \times \mathbb{C}^8 \times \mathbb{C}^8 \times \mathbb{C},$$

$$z_{23} = (y_1, \ldots, y_8),$$

$$C_1(z) = \begin{pmatrix} s_1 & z_{12} & z_{13} \\ z'_{12} & s_2 I^{(8)} & R(z_{23}) \\ z'_{13} & R(z_{23})' & s_3 I^{(8)} \end{pmatrix},$$

$$C_2(z) = \begin{pmatrix} s_2 & z_{23} \\ z'_{23} & s_3 I^{(8)} \end{pmatrix},$$

$$C_3(z) = s_3,$$

$$R(z_{23}) = \sum_r e'_r z_{23} (A_{12}^{r3})'$$

$$\text{(a)} \quad = \begin{pmatrix} y_1 & y_2 & y_3 & y_4 & y_5 & y_6 & y_7 & y_8 \\ -y_2 & y_1 & y_4 & -y_3 & y_6 & -y_5 & -y_8 & y_7 \\ -y_3 & -y_4 & y_1 & y_2 & y_7 & y_8 & -y_5 & -y_6 \\ -y_4 & y_3 & -y_2 & y_1 & y_8 & -y_7 & y_6 & -y_5 \\ -y_5 & -y_6 & -y_7 & -y_8 & y_1 & y_2 & y_3 & y_4 \\ -y_6 & y_5 & -y_8 & y_7 & -y_2 & y_1 & -y_4 & y_3 \\ -y_7 & y_8 & y_5 & -y_6 & -y_3 & y_4 & y_1 & -y_2 \\ -y_8 & -y_7 & y_6 & y_5 & -y_4 & -y_3 & y_2 & y_1 \end{pmatrix}$$

and

$$x = (r_1, y_{12}, r_2, y_{13}, y_{23}, r_3) \in \mathbb{R} \times \mathbb{R}^8 \times \mathbb{R} \times \mathbb{R}^8 \times \mathbb{R}^8 \times \mathbb{R},$$

$$\Delta(x) = r_1 r_2 r_3 - r_3|y_{12}|^2 - r_2|y_{13}|^2 - r_1|y_{23}|^2 + 2y_{12}R(y_{23})y_{13}'.$$

Using Theorem 2.1 and noticing that $\mu_1 = -18$, $\mu_2 = 126$, $\mu_3 = -882$, the Bergman kernel of $R_{VI}(27)$ becomes

$$K(z,\bar{z}) = c_0(\det C_1(\mathrm{Im}(z)))^{-18}(\det C_2(\mathrm{Im}(z)))^{126}(\det C_3(\mathrm{Im}(z)))^{-882}$$
$$= c_0 \Delta(\mathrm{Im}(z))^{-9}.$$

In this case, $\nu_j = \frac{1}{2}\mu_j$, $j = 1,2,3$, the Cauchy-Szego kernel becomes

$$S(z,\xi) = c_1 \Delta \left(\frac{1}{2\sqrt{-1}}(z - \xi) \right)^{-9}, \qquad \forall z \in R_{VI}(27), \xi \in \mathbb{R}^{27}.$$

It follows that the Poisson kernel of $R_{VI}(27)$ has the form

$$P(z,\xi) = |S(z,\xi)|^2 / S(z,\bar{z}) = c_1 \Delta(\mathrm{Im}(z))^9 \left| \Delta \left(\frac{1}{2\sqrt{-1}}(z - \xi) \right) \right|^{-18},$$

where $z \in R_{VI}(27)$, $\xi \in \mathbb{R}^{27}$.

The holomorphic automorphism group of $R_{VI}(27)$ is

$$\mathrm{Aut}(R_{VI}(27)) = \mathrm{Iso}_{(\sqrt{-1}v_0, 0)}(R_{VI}(27))T(R_{VI}(27)),$$

where $T(R_{VI}(27))$ is a closed subgroup of $\mathrm{Aut}(R_{VI}(27))$, which acts simply transitively on $R_{VI}(27)$. Let $t(R_{VI}(27))$ be the Lie algebra of $T(R_{VI}(27))$. Then $t(R_{VI}(27))$ has a basis

$$\frac{\partial}{\partial s_i}, \quad 1 \leqslant i \leqslant 3, \qquad \frac{\partial}{\partial z_{ij}^t}, \quad 1 \leqslant t \leqslant 8, 1 \leqslant i < j \leqslant 3,$$

$$A_1 = 2s_1 \frac{\partial}{\partial s_1} + z_{12} \frac{\partial'}{\partial z_{12}} + z_{13} \frac{\partial'}{\partial z_{13}},$$

$$A_2 = 2s_2 \frac{\partial}{\partial s_2} + z_{12} \frac{\partial'}{\partial z_{12}} + z_{23} \frac{\partial'}{\partial z_{23}},$$

$$A_3 = 2s_3 \frac{\partial}{\partial s_3} + z_{13} \frac{\partial'}{\partial z_{13}} + z_{23} \frac{\partial'}{\partial z_{23}},$$

$$X_{12}^t = 2z_{12}^t \frac{\partial}{\partial s_1} + s_2 \frac{\partial}{\partial z_{12}^t} + z_{23}(A_{12}^{t3})' \frac{\partial'}{\partial z_{13}},$$

$$X_{13}^t = 2z_{13}^t \frac{\partial}{\partial s_1} + s_3 \frac{\partial}{\partial z_{13}^t} + \sum z_{23}(A_{12}^{r3})' e_t' \frac{\partial}{\partial z_{12}^r},$$

$$X_{23}^t = 2z_{23}^t \frac{\partial}{\partial s_2} + s_3 \frac{\partial}{\partial z_{23}^t} + \sum z_{13} A_{12}^{r3} e_t' \frac{\partial}{\partial z_{12}^r}.$$

We now introduce the holomorphic isomorphism σ by defining

$$r_1 = 3\sqrt{-2} + 6\sqrt{2}\Delta_0^{-1} \left[(s_2 + \sqrt{-1})(s_3 + \sqrt{-1}) - z_{23}z_{23}' \right],$$

$$r_2 = 3\sqrt{-2} + 6\sqrt{2}\Delta_0^{-1} \left[(s_1 + \sqrt{-1})(s_3 + \sqrt{-1}) - z_{13}z_{13}' \right],$$

$$r_3 = 3\sqrt{-2} + 6\sqrt{2}\Delta_0^{-1} \left[(s_1 + \sqrt{-1})(s_2 + \sqrt{-1}) - z_{12}z_{12}' \right],$$

$$x_{12} = 12\Delta_0^{-1} \left[\sum (z_{13}A_{12}^{r3}z_{23}')e_r - (s_3 + \sqrt{-1})z_{12} \right],$$

$$x_{13} = 12\Delta_0^{-1} \left[\sum z_{12}^r z_{23}(A_{12}^{r3})' - (s_2 + \sqrt{-1})z_{13} \right],$$

$$x_{23} = 12\Delta_0^{-1} \left[\sum z_{12}^r z_{13}A_{12}^{r3} - (s_1 + \sqrt{-1})z_{23} \right],$$

where

$$\Delta_0 = (s_1 + \sqrt{-1})(s_2 + \sqrt{-1})(s_3 + \sqrt{-1}) - (s_1 + \sqrt{-1})z_{23}z_{23}'$$
$$- (s_2 + \sqrt{-1})z_{13}z_{13}' - (s_3 + \sqrt{-1})z_{12}z_{12}' + 2\sum z_{12}^r z_{13}A_{12}^{r3}z_{23}',$$

and $\sigma(\sqrt{-1}v_0) = 0$. Let $\mathfrak{iso}_0(\sigma(R_{VI}(27)))$ be the Lie algebra of $\mathrm{Iso}_0(\sigma(R_{VI}(27)))$. Then it has a basis

$$\{B_1, B_2, B_3, P_{jk}, 1 \leqslant j < k \leqslant 8, Z_{12}(\alpha), Z_{13}(\beta), Z_{23}(\gamma)\},$$

where

$$B_1 = \sqrt{-1} \left(2r_1 \frac{\partial}{\partial r_1} + x_{12} \frac{\partial'}{\partial x_{12}} + x_{13} \frac{\partial'}{\partial x_{13}} \right),$$

$$B_2 = \sqrt{-1} \left(2r_2 \frac{\partial}{\partial r_2} + x_{12} \frac{\partial'}{\partial x_{12}} + x_{23} \frac{\partial'}{\partial x_{23}} \right),$$

$$B_3 = \sqrt{-1} \left(2r_3 \frac{\partial}{\partial r_3} + x_{13} \frac{\partial'}{\partial x_{13}} + x_{23} \frac{\partial'}{\partial x_{23}} \right),$$

$$P_{jk} = 2 \left(x_{12}^j \frac{\partial}{\partial x_{12}^k} - x_{12}^k \frac{\partial}{\partial x_{12}^j} \right) - x_{13}P_j P_k \frac{\partial'}{\partial x_{13}} + x_{23}P_k P_j \frac{\partial'}{\partial x_{23}},$$

$$1 \leqslant j < k \leqslant 8,$$

$$Z_{12}(\alpha) = \alpha x_{12}' \frac{\partial}{\partial r_1} - \bar{\alpha}x_{12}' \frac{\partial}{\partial r_2} + r_2\alpha \frac{\partial'}{\partial x_{12}} - r_1\bar{\alpha} \frac{\partial'}{\partial x_{12}}$$

$$+ \frac{1}{\sqrt{2}} \sum \alpha_t x_{23}P_t' \frac{\partial'}{\partial x_{13}} - \frac{1}{\sqrt{2}} \sum \bar{\alpha}_t x_{13}P_t \frac{\partial'}{\partial x_{23}},$$

$$Z_{13}(\beta) = \beta x_{13}' \frac{\partial}{\partial r_1} - \bar{\beta}x_{13}' \frac{\partial}{\partial r_3} + r_3\beta \frac{\partial'}{\partial x_{13}} - r_1\bar{\beta} \frac{\partial'}{\partial x_{13}}$$

$$+\frac{1}{\sqrt{2}}\sum \beta P_r x_{23}' \frac{\partial'}{\partial x_{12}^r} - \frac{1}{\sqrt{2}}\sum x_{12}^r \bar{\beta} P_r \frac{\partial'}{\partial x_{23}},$$

$$Z_{23}(\gamma) = \gamma x_{23}' \frac{\partial}{\partial r_2} - \bar{\gamma} x_{23}' \frac{\partial}{\partial r_3} + r_3 \gamma \frac{\partial'}{\partial x_{23}} - r_1 \bar{\gamma} \frac{\partial'}{\partial x_{23}}$$

$$+\frac{1}{\sqrt{2}}\sum x_{13} P_r \gamma' \frac{\partial'}{\partial x_{12}^r} - \frac{1}{\sqrt{2}}\sum x_{12}^r \bar{\gamma} P_r' \frac{\partial'}{\partial x_{13}},$$

and $P_t = A_{12}^{t3}$, $1 \leqslant t \leqslant 8$, $\alpha, \beta, \gamma \in \mathbb{C}^8$.

4. Holomorphic Automorphism Group of N-Siegel Domains

The holomorphic automorphism group for homogeneous bounded domains had been considered by a lot of mathematicians. First, L. K. Hua studied it in the classical domains of the first four classes. Then, Kaup, Matsushima and Ochiai [6] obtained further properties. In 1975, Dorfmeister [2] derived it by using the non-associated algebras for homogeneous Siegel domains.

We have determined completely the maximal connected holomorphic automorphism group of N-Siegel domains in 1976 [15].

THEOREM 4.1 [15]. *Let $D(V_N, F)$ be an N-Siegel domain defined by the N-complex matrix set $\{A_{ij}^{tk}, Q_{ij}^t\}$ and let $\mathrm{Aut}(D(V_N, F))$ be the holomorphic automorphism group of $D(V_N, F)$. Let $\mathfrak{aut}(D(V_N, F))$ be the Lie algebra of $\mathrm{Aut}(D(V_N, F))$. Then*

$$\mathfrak{aut}(D(V_N, F)) = L_{-2} + L_{-1} + L_0 + L_1 + L_2,$$

where

$$L_{-2} = \langle \frac{\partial}{\partial s_j}, 1 \leqslant j \leqslant N, \frac{\partial}{\partial z_{ij}^t}, 1 \leqslant t \leqslant n_{ij}, 1 \leqslant i < j \leqslant N \rangle,$$

$$L_{-1} = \left\{ \alpha \frac{\partial'}{\partial u} + 2\sqrt{-1} F(u, \alpha) \frac{\partial'}{\partial z} \ \middle| \ \forall \alpha \in \mathbb{C}^m \right\}$$

$$L_0 = \langle A_j, 1 \leqslant j \leqslant N, X_{ij}^t, 1 \leqslant t \leqslant n_{ij}, 1 \leqslant i < j \leqslant N \rangle + L_{01} + L_{02}$$

$$A_j = 2s_j \frac{\partial}{\partial s_j} + \sum_{\ell < j} z_{\ell j} \frac{\partial'}{\partial z_{\ell j}} + \sum_{j < \ell} z_{j\ell} \frac{\partial'}{\partial z_{j\ell}} + u_j \frac{\partial'}{\partial u_j},$$

$$X_{ij}^t = 2z_{ij}^t \frac{\partial}{\partial s_i} + s_j \frac{\partial}{\partial z_{ij}^t} + \sum_{\ell < j} \sum_r (z_{\ell j} A_{\ell i}^{rj} e_t') \frac{\partial}{\partial z_{\ell i}^r}$$

$$+ \sum_{i < \ell < j} \sum_r (e_t A_{i\ell}^{rj} z_{\ell j}') \frac{\partial'}{\partial z_{i\ell}} + \sum_{j < \ell} z_{j\ell} (A_{ij}^{t\ell})' \frac{\partial'}{\partial z_{i\ell}} + u_j (\overline{Q_{ij}^t})' \frac{\partial'}{\partial u_i},$$

and

$$L_{01} = \left\{ \sum_{i < j} z_{ij} L_{ij} \frac{\partial'}{\partial z_{ij}} + \sum_i u_i K_i \frac{\partial'}{\partial u_i} \ \middle| \ \bar{L}_{ij} = L_{ij} = -L_{ij}', K_i + \bar{K}_i' = 0, \right.$$

$$\left. L_{ik} A_{ij}^{tk} - A_{ij}^{tk} L_{jk} = \sum_r (e_r L_{ij} e_t') A_{ij}^{rk}, K_i Q_{ij}^t - Q_{ij}^t K_j = \sum_r (e_r L_{ij} e_t') Q_{ij}^r \right\}$$

$Z_{ij}^t \in L_{02}$, $1 \leqslant t \leqslant n_{ij}$ if and only if $n_{ij} > 0$ and when $\ell \in \{i+1, \ldots, j-1\}$ such that $n_{i\ell} > 0$, then

$$n_{i\ell} = n_{\ell j} = n_{ij}, \qquad m_i = m_\ell = m_j,$$

where n_{ij} is equal to one of the integers 1, 2, 4, 8, and

$$n_{ki} = n_{k\ell} = n_{kj}, \qquad 1 \leqslant k \leqslant i, \qquad n_{ik} = n_{k\ell} = n_{kj}, \qquad 1 < k < \ell,$$

$$n_{ik} = n_{\ell k} = n_{kj}, \qquad \ell < k < j, \qquad n_{ik} = n_{\ell k} = n_{jk}, \qquad j < k,$$

$$Z_{ij}^t = 2z_{ij}^t \frac{\partial}{\partial s_j} + s_i \frac{\partial}{\partial z_{ij}^t} + \sum_{\ell < i} \sum_r z_{\ell i}^r e_t (A_{\ell i}^{rj})' \frac{\partial'}{\partial z_{\ell j}} + \sum_{i < \ell < j} \sum_r z_{i\ell}^r e_t A_{i\ell}^{rj} \frac{\partial'}{\partial z_{\ell j}}$$

$$+ \sum_{j < \ell} z_{i\ell} A_{ij}^{t\ell} \frac{\partial'}{\partial z_{j\ell}} + u_i Q_{ij}^t \frac{\partial'}{\partial u_j}.$$

If in the above equation for all indices $1 \leqslant i_1 < \cdots < i_\sigma \leqslant N$,

$$n_{i_\rho \ell} = 0, \qquad i_\rho < \ell, \qquad \ell \neq i_{\rho+1}, \ldots, i_\sigma,$$

$$Z_{i_\rho i_\eta}^t \in L_{02}, \qquad 1 \leqslant t \leqslant n_{i_\rho i_\eta}, \qquad 1 \leqslant \rho < \eta \leqslant \sigma,$$

and

$$\sum_t Q_{\ell i_\rho}^t (e_u' e_v + e_v' e_u)(Q_{\ell i_\rho}^t)' = 0, \qquad 1 \leqslant u, v \leqslant m_{i_\rho}, 1 \leqslant \ell < i_\rho, 1 \leqslant \rho \leqslant \sigma;$$

$$\sum_t (Q_{\ell i_\rho}^t)'(e_u' e_v + e_v' e_u) Q_{i_\rho \ell}^t = 0, \qquad 1 \leqslant u, v \leqslant m_{i_\rho}, i_\rho < \ell \leqslant N, 1 \leqslant \rho \leqslant \sigma,$$

then

$$L_1 = \langle Y_{i_\rho}^t, \tilde{Y}_{i_\rho}^t, 1 \leqslant t \leqslant m_{i_\rho}, 1 \leqslant \rho \leqslant \sigma \rangle,$$

$$L_2 = \langle B_{i_\rho}, 1 \leqslant \rho \leqslant \sigma, T_{i_\rho i_\eta}^t, 1 \leqslant t \leqslant n_{i_\rho i_\eta}, 1 \leqslant \rho < \eta \leqslant \sigma \rangle,$$

where

$$B_j = s_j \left(A_j + u_j \frac{\partial'}{\partial u_j} \right) + \sum_{\ell < j} \sum_r z_{\ell j}^r \left(X_{\ell j}^r + u_j (\overline{Q_{\ell j}^r})' \frac{\partial'}{\partial u_\ell} \right)$$

$$+ \sum_{j < \ell} \sum_r z_{j\ell}^r \left(Z_{j\ell}^r + u_j Q_{j\ell}^r \frac{\partial'}{\partial u_\ell} \right),$$

$$T_{ij}^t = [X_{ij}^t, B_i].$$

And these conditions imply that when $n_{\ell i_\rho} > 0$ and $Z_{\ell i_\rho}^t \in L_{02}$, then ℓ is one of the indices $i_1, \ldots, i_{\rho-1}$.

COROLLARY 1. *Let* $\mathrm{Aff}(D(V_N, F))$ *be the linear transformation group of the N-Siegel domain* $D(V_N, F)$ *and* $\mathfrak{aff}(D(V_N, F))$ *the Lie algebra of* $\mathrm{Aff}(D(V_N, F))$. *Then*

$$\mathfrak{aff}(D(V_N, F)) = L_{-2} + L_{-1} + L_0.$$

Using the expression of Bergman kernel for *N*-Siegel domains, we have

THEOREM 4.2. *Let $D(V_N, F)$ be an N-Siegel domain. Then the Bergman mapping σ:*

$$(z, u) \to \nabla_{(\bar{z}, \bar{u})} \log \frac{K(z, u; \bar{z}, \bar{u})}{K(\sqrt{-1}v_0, 0; \bar{z}, \bar{u})}\Bigg|_{\bar{z} = -\sqrt{-1}v_0, \bar{u} = 0} \quad T(\sqrt{-1}v_0, 0; -\sqrt{-1}v_0, 0)^{-\frac{1}{2}}$$

is a holomorphic isomorphism on $D(V_N, F)$, where $v_0 = (1, 0, 1, \ldots, 0, 1) \in V_N$, ∇ is the gradient of the complex Euclidean metric, and $\sigma(D(V_N, F))$ is a homogeneous bounded domain.

COROLLARY 2. *Let σ be the Bergman mapping of an N-Siegel domain $D(V_N, F)$. Then $\mathrm{Aff}(\sigma(D(V_N, F)))$ is the isotropic group at a fixed point $\sigma(\sqrt{-1}v_0, 0) = (0, 0)$.*

5. The Classification of Square Cones and Dual Square Cones

LEMMA 5.1. *Let $A_1, \ldots, A_n$ be $m \times m$ real matrices satisfying the condition*

$$A_i' A_j + A_j' A_i = 2\delta_{ij} I, \qquad 1 \leqslant i, j \leqslant n.$$

Then the canonical form for the equivalent relation

$$A_j \to \sum_{r=1}^{n} (e_r O e_j') O_1 A_r O_2, \qquad 1 \leqslant j \leqslant n$$

is $\{P_1, \ldots, P_n\}$, where $m = 2^{[\frac{n-1}{2}]}M$ and $O \in O(n)$, $O_1, O_2 \in O(m)$,

$$P_1 = I,$$
$$P_{2j+1} = \begin{pmatrix} BR_{2j-1} & 0 \\ 0 & -BR_{2j-1} \end{pmatrix},$$
$$P_2 = \begin{pmatrix} 0 & -I_1 \\ I_1 & 0 \end{pmatrix},$$
$$P_{2j+2} = \begin{pmatrix} 0 & -\sqrt{-1}BR_{2j} \\ -\sqrt{-1}BR_{2j} & 0 \end{pmatrix}$$

$j = 1, 2, \ldots$. *Here $R_1, \ldots, R_{n-2}$ are the canonical forms under the equivalent relation*

$$Q_j = \sum_r (e_r O e_j') U Q_r V, \qquad \bar{Q}_j' Q_k + \bar{Q}_k' Q_j = 2\delta_{jk} I,$$

for the order $m_1 = m/2$:

$$R_1 = I,$$
$$R_{2j} = \sqrt{-1} \begin{pmatrix} & & \begin{pmatrix} I_j & 0 \\ 0 & -I_j \end{pmatrix} \\ & \cdot^{\cdot^{\cdot}} & \\ \begin{pmatrix} I_j & 0 \\ 0 & -I_j \end{pmatrix} & & \end{pmatrix},$$

$$R_{2j+1} = \begin{pmatrix} & & & \begin{pmatrix} 0 & I_j \\ -I_j & 0 \end{pmatrix} \\ & \cdot^{\cdot^{\cdot}} & \\ \begin{pmatrix} 0 & I_j \\ -I_j & 0 \end{pmatrix} & & \end{pmatrix},$$

$j = 1, 2 \ldots, [\frac{n-3}{2}]$, and when n is an even number,

$$R_{n-2} = \sqrt{-1} \begin{pmatrix} & & & \begin{pmatrix} I^{(p)} & 0 \\ 0 & -I^{(q)} \end{pmatrix} \\ & \cdot^{\cdot^{\cdot}} & \\ \begin{pmatrix} I^{(p)} & 0 \\ 0 & -I^{(q)} \end{pmatrix} & & \end{pmatrix}$$

and $p + q = M$, $p \geqslant 0$, $q \geqslant 0$,

$$B_1 = B, \qquad B_k = \begin{pmatrix} S_k & 0 \\ 0 & -S_k \end{pmatrix}, \qquad S_k = \begin{pmatrix} 0 & -B_{k+1} \\ B_{k+1} & 0 \end{pmatrix},$$

$k = 1, 2, \ldots$, the blocks B_k, S_k with the least order is

(1) when $n = 8k - 3$, $B_{2k} = I$;
(2) when $n = 8k - 2$, $B_{2k} = I$, and so $p = q$;
(3) when $n = 8k - 1$, $B_{2k} = I$;
(4) when $n = 8k$, $B_{2k} = I$, where $p + q = M$ and $p \geqslant q$;
(5) when $n = 8k + 1$, $S_{2k} = I$;
(6) when $n = 8k + 2$,

$$S_{2k} = \begin{pmatrix} 0 & I \\ I & 0 \end{pmatrix},$$

and so $p = q$;
(7) when $n = 8k + 3$,

$$B_{2k+1} = \begin{pmatrix} 0 & -I \\ I & 0 \end{pmatrix};$$

(8) when $n = 8k + 4$,

$$B_{2k+1} = \begin{pmatrix} \begin{pmatrix} 0 & -I \\ I & 0 \end{pmatrix}^{(p)} & 0 \\ 0 & \begin{pmatrix} 0 & -I \\ I & 0 \end{pmatrix}^{(q)} \end{pmatrix};$$

and moreover in this case the integers p and q both are even numbers and $p \geqslant q$.

In all the above cases, the positive integers n and p are the total invariant.

COROLLARY 1. *Let* $A_0, \ldots, A_n$, $n \geqslant 1$ *be* $n + 1$ *real symmetric matrices of order* m. *Suppose that they satisfy the conditions*

$$A_i A_j + A_j A_i = 2\delta_{ij} I, \qquad 0 \leqslant i, j \leqslant n.$$

Then the canonical form for the equivalent relation

$$A_i \rightarrow \sum_{r=1}^{n} (e_r O e_i') O_1 A_r O_1', \qquad 0 \leqslant i \leqslant n$$

is $\{D_1, \ldots, D_n\}$, *where* $O \in O(n+1)$, $O_1 \in O(m)$ *and*

$$D_0 = \begin{pmatrix} I_1 & 0 \\ 0 & -I_1 \end{pmatrix}, \qquad D_i = \begin{pmatrix} 0 & P_i \\ P_i' & 0 \end{pmatrix}, \qquad 1 \leqslant i \leqslant n$$

where $\{P_1, \ldots, P_n\}$ *is defined by Lemma 5.1. Moreover a total invariant is*

$$\operatorname{tr} A_0 A_1 \ldots A_n.$$

DEFINITION. *If* N-*cones* V_N *and* $\tilde{V}_N$ *are defined by the* N-*matrix sets* $\{A_{ij}^{tk}\}$ *and* $\{B_{ij}^{tk}\}$ *respectively, where*

$$B_{ij}^{tk} = \sum_r A_{N+1-k,N+1-j}^{r,N+1-i} e_t' e_r, \qquad \tilde{n}_{ij} = n_{N+1-j,N+1-i},$$

then the N-*cone* $\tilde{V}_N$ *is said to be the adjoint cone of the* N-*cone* V_N.

LEMMA 5.2. *If an* N-*cone* V_N *is defined by the matrix set* $\{A_{ij}^{tk}\}$, *then the adjoint* N-*cone* $\tilde{V}_N$ *of the* N-*cone* V_N *has the matrix representation*

$$\tilde{R}_j(x) =$$
$$\begin{pmatrix}
r_1 I & \sum_t x_{12}^t A_{12}^{tj} & \cdots & \sum_t x_{1,j-1}^t A_{i,j-1}^{tj} & x_{1j}' \\
\sum_t x_{12}^t (A_{12}^{tj})' & r_2 I & \cdots & \sum_t x_{2,j-1}^t A_{2,j-1}^{tj} & x_{2j}' \\
\vdots & \vdots & \ddots & \vdots & \vdots \\
\sum_t x_{1,j-1}^t (A_{1,j-1}^{tj})' & \sum_t x_{2,j-1}^t (A_{2,j-1}^{tj})' & \cdots & r_{j-1} I & x_{j-1,j}' \\
x_{1j}' & x_{2j} & \cdots & x_{j-1,j} & r_j
\end{pmatrix}$$
$$> 0.$$

$i = 1, 2, \ldots, N$, *where* $r_i = \tilde{r}_{N+1-i}$, $x_{ij} = \tilde{x}_{N+1-j,N+1-i}$ *and* $\tilde{x} = (\tilde{r}_1, \tilde{x}_2, \tilde{r}_2, \ldots, \tilde{x}_N, \tilde{r}_N)$. ·

LEMMA 5.3. *If a homogeneous cone* V *is linearly equivalent to the* N-*cone* V_N, *then the dual cone* V^* *of the cone* V *is linearly equivalent to the adjoint cone* $\tilde{V}_N$ *of the* N-*cone* V_N. *Moreover if* $\tilde{V}_N^{(1)}$ *and* $\tilde{V}_N^{(2)}$ *are adjoint cones of the* N-*cones* $V_N^{(1)}$ *and* $V_N^{(2)}$ *respectively, then* $V_N^{(1)}$ *is linearly equivalent to* $V_N^{(2)}$ *if and only if* $\tilde{V}_N^{(1)}$ *is linearly equivalent to* $\tilde{V}_N^{(2)}$.

LEMMA 5.4. *An* N-*cone* V_N *is a cone of the square type if and only if its adjoint cone* $\tilde{V}_N$ *is an* N-*cone of the dual square type.*

COROLLARY 1 (KOECHER-VINBERG). *Any linearly homogeneous cone of the self-dual is linearly equivalent to the topological product of canonical cones defined below:*

(1) $V_1 = \{r \in \mathbb{R} \mid r > 0\}$;

(2) $V_2 = V_2(\sigma_2) = \{(r_1, x_{12}, r_2) \in \mathbb{R} \times \mathbb{R}^{n_{12}} \times \mathbb{R} \mid r_2 > 0, r_1 r_2 - x_{12} x'_{12} > 0\}$;

(3) $V_3(8,8) = \{(r_1, x_{12}, r_2, x_{13}, x_{23}, r_3) \in \mathbb{R} \times \mathbb{R}^8 \times \mathbb{R} \times \mathbb{R}^8 \times \mathbb{R}^8 \times \mathbb{R} \mid$

$$
\begin{pmatrix}
r_1 & x_{12} & x_{13} \\
x'_{12} & r_2 I & R_{23} \\
x'_{13} & R'_{23} & r_3 I
\end{pmatrix} > 0, \}
$$

where $R_{23} = R(x_{23})$ is defined by (a) *in section 3;*

(4) $V_N(1,1,\ldots,1) = \{S > 0\}$, *where S is a real symmetric matrix of order N;*

(5) $V_N(2,2,\ldots,2) = \{H > 0\}$, *where H is a Hermitian matrix of order N;*

(6) $V_N(4,4,\ldots,4) = \{Q > 0\}$, *where Q is a quarternion Hermitian matrix of order N.*

LEMMA 5.5. *N-cone V_N is linearly equivalent to one of V_1, $V_2(\sigma_2)$, $V_3(8,8)$, $V_N(1,1,\ldots,1)$, $V_N(2,2,\ldots,2)$, $V_N(4,4,\ldots,4)$ if and only if V_N is a square N-cone and V_N is linearly equivalent to the dual square N-cone.*

6. The Classification of N-Siegel Domains over the Square Cones and Dual Square Cones

As we know that the classification of the square cones and the dual square cones, in the sense of linearly equivalence, are identical with that of the tube domains over these cones. Thus from section 5 we have a complete classification of N-Siegel domains for tube domains over these cones.

DEFINITION. Let $D(V_N, F)$ be an N-Siegel domain, where V_N is a square N-cone or a dual square N-cone. Then $D(V_N, F)$ is said to be a square domain or a dual square domain respectively.

The classification of certain square domains and dual square domains were obtained in [18]. Particularly, when $1 \leqslant \tau < N$, $m_1 = \cdots = m_\tau > m_{\tau+1} = \cdots = m_N = 0$, we have characterized all the canonical domains.

REFERENCES

1. Cartan, E., *Sur les domaines bornes homogenes de l'espace de n variables complexes*, Abh. Math. Sem. **11** (1935), Hamburg, 116–162.

2. Dorfmeister, J., *Homogene Siegel Gebiete*, Habilitation (197), Univ. Munster.

3. Gindikin, S. G., *Analysis in homogeneous domains*, Uspehi Mat., Nauk **19** (1964), 3–92; Russian Math., Survefs **19** (1964).

4. Hua, L. K., *Harmonic Analysis of Functions of Several Complex Variables in the Classical Domains*, Transl. Math. Mono., vol. 6, Amer. Math. Soc., Providence, 1963.

5. Keneyuki, S., *Homogeneous bounded domains and Siegel domains*, Lecture Notes in Mathematics, No. 241 (1971), Springer.

120 YICHAO XU

6. Kaup, W., Matsushima, Y. and Ochiai, T., *On the automorphisms and equivalences of generalized Siegel domains*, Amer. J. Math. **92** (1970), 475–497.
7. Koranyi, A., *The Poisson integral for generalized half-planes and bounded symmetric domains*, Ann. Math. **82** (1965), 332–350.
8. Look, K. H. and Xu Yichao (Hsu I Chao), *A note on transitive domains*, Acta Math. Sinica **11** (1961), 11–23; Chinese Math. **2** (1962), 11–26.
9. Murakami, S., *On Automorphisms of Siegel domains*, Lecture Notes in Mathematics, No. 286, Springer, 1972.
10. Pjatetski-Shapiro, I. I., *Geometry of Classical Domains and Theory of Automorphic Functions*, Gordon and Breach, New York, 1969.
11. Vagi, S., *Harmonic analysis on Cartan and Siegel domains*, Studies in Math. DePaul Univ. **13** (1976), 257–309.
12. Vinberg, E. B., *The theory of convex homogeneous cones*, Trans. Moscow Math. Soc. (1963), 340–403.
13. Vinberg, Gindikin and Pjatetski-Shapiro, *Classification and canonical realization of complex bounded homogeneous domains*, Trans. Moscow Math. Soc. (1963), 404–437.
14. Xu Yichao (Hsu I Chao), *On the classification of symmetric schlicht domains in several complex variables*, Shuxue Jinzhan **8** (1965), 109–144.
15. Xu Yichao, *On the automorphism group of the homogeneous bounded domains*, Acta Math. Sinica **19** (1976), 169–191.
16. Xu Yichao, *On the isomorphisms of the homogeneous bounded domains*, Acta Math. Sinica **20** (1977), 248–266.
17. Xu Yichao, *The Siegel domains of first kind associated with the cones of square type*, Acta Math. Sinica **21** (1978), 1–17.
18. Xu Yichao, *Classification of square type domains*, Scientia Sinica **22** (1979), 375–392.
19. Xu Yichao, *On the Bergman kernel function of homogeneous bounded domains*, Scientia Sinica, Special Issue **2** (1979), 80–90.
20. Xu Yichao, *A note on the homogeneous Siegel domains*, Acta Math. Sinica **24** (1981), 99–105.
21. Xu Yichao, *Tube domains over cones with dual square type*, Scientia Sinica **24** (1981), 1475–1488.
22. Xu Yichao, *Some results on the classification of homogeneous bounded domains*, Proceedings of the 1980 Beijing Symposium on Differential Geometry and Differential Equations, vol. 3, Gordon and Breach Sci. Pub. Inc., 1981, pp. 1621–1637.
23. Xu Yichao, *The canonical realization of homogeneous bounded domains*, Scientia Sinica **26** (1983), 25–34.
24. Xu Yichao and Wu Lanfang, *The holomorphic sectional curvature of homogeneous bounded domains*, Kexue Tongbao **28** (1983), 592–596.
25. Xu Yichao, *Classification of a class of homogeneous Kählerian Manifolds*, Scientia Sinica (Series A) **29** (1986), 449–463.

INSTITUTE OF MATHEMATICS, ACADEMIA SINICA, BEIJING 100080, PEOPLE'S REPUBLIC OF CHINA

Contemporary Mathematics
Volume **142**, 1993

Differential Operators and Function Spaces

ZHIMIN YAN

In this paper, we shall consider some aspects of analysis on symmetric cones and bounded symmetric domains, and their relations as well. In particilar, we are interested in studying the algebra $D(\Omega)$ of the invariant differential operators on a symmetric cone Ω. We will give some sets of generators and calculate the eigenvalues of spherical functions under those generators. The explicit construction of our invariant differential operators in $D(\Omega)$ leads to introducing some differential operaotrs on a bounded symmetric domain D. These differential operators will be used in our study of spaces of holomorphic functions on the bounded symmetric domain D, which occur in representation theory as well as function theory. Another subject that concerns us is about the generalized hypergeometric functions associated with Ω. These special functions have appeared in statistics, harmonic analysis and other fields in mathematics. Here we are mainly concerned with those aspects related to complex analysis on bounded symmetric domains. In particular, the generalized Gaussian hypergeometric function will be characterized by a system of differential equations which are closely related to Hua-equations. The nature of this paper is descriptive and expository, and the details will appear somewhere else.

§1. Invariant differential operators and generalized hypergeometric functions associated with symmetric cones

In this part, we shall consider analysis on a symmetric cone Ω. First, we shall study the algebra $D(\Omega)$ of the invariant differential operators on Ω. It is known, for instance, see [**H**], that the algebra of the invariant differential operators on a symmetric space of rank r is commutative and generated by a set of r algebraically independent elements. An important problem is to give such a set of generators explicitly. For this purpose, corresponding to each complex number

1991 *Mathematics Subject Classification.* 32A07, 32A35, 32A40, 32M15, 33C70, 43A85, 46B10, 46E15, 46E20.

The final version of this paper will be submitted for publication elsewhere.

λ, an invariant differential operator D_λ is introduced. It is shown that for any r distinct numbers $\lambda_1, \ldots, \lambda_r$, $D_{\lambda_1}, \ldots, D_{\lambda_r}$ is a set of algebraically independent generators of $D(\Omega)$. We also introduce r "canonical" invariant differential operators K_j which can be constructed naturally from some canonical invariant polynomials, then express D_λ in terms of $K_1, \ldots, K_r$. An important feature of those invariant differential operators is that for any spherical function, its eigenvalues under D_λ and K_j can be computed explicitly.

Secondly, we will consider a class of generalized hypergeometric functions $_pF_q$ associated with the cone Ω. One will see that $_2F_1$, the generalized Gaussian hypergeometric function, is the unique analytic solution of a system of generalized hypergeometric equations subject to certain symmetries and a normalization at the origin. These equations are closely connected with the Hua-equations studied in [**JK, L**]. We will also discuss the asymptotic behavior of the hypergeometric function $_{q+1}F_q$ whose applications will be given in the second part of this paper.

§1.1. Background and notation of symmetric cones and Jordan algebra

We refer to [**F**] for further detail about this section.

Let V be a real simple Euclidean Jordan algebra, then the interior Ω of the set of all squares in V is a symmetric cone in V. Conversely, every irreducible symmetric cone can be realized as the interior of the set of all squares in a simple Euclidean Jordan algebra.

We fix a Jordan frame $\{c_1, \ldots, c_r\}$ of V where r is the rank of V, then the identity element e is equal to $c_1 + \cdots + c_r$.

We denote by $G(\Omega)$ the identity component of the subgroup of $GL(V)$ which preserves Ω, L the isotropy subgroup of $G(\Omega)$ at e. Then every element x in V can be written as

$$(1) \qquad x = l. \sum_{i=1}^{r} t_i c_i, \qquad t_i \in \mathbb{R}, \quad l \in L,$$

and $x \in \Omega$ if and only if $t_i > 0$, $i = 1, \ldots, r$.

There is a determinant polynomial $\Delta(x)$ on V, which is L-invariant, such that if x is written as in (1), then

$$\Delta(x) = \prod_{i=1}^{r} t_i.$$

For an idempotent c and $k \in \mathbb{R}$, let

$$V(c, k) = \{x \in V \mid L(c)x = kx\},$$

and $V_{ij} = V(c_i, 1/2) \cap V(c_j, 1/2)$ for $i \neq j$. Then all V_{ij} have the same dimension, which will be denoted by a.

Now the subspaces $V^{(k)} = V(c_1 + \cdots + c_k, 1)$ $(1 \leqslant k \leqslant r)$ are subalgebras of V. Let P_k be the orthogonal projection onto $V^{(k)}$. The principal minor $\Delta_k(x)$ is the polynomial defined on V by

$$\Delta_k(x) = \Delta^{(k)}(P_k x),$$

where $\Delta^{(k)}$ is the determinant polynomial with respect to the algebra $V^{(k)}$. The $\Delta_k(x)$ are positive if $x \in \Omega$.

When V is the space of real symmetric matrices or Hermitian matrices, then Δ_k are the ordinary principal minors.

For an r-tuple of integers $\mathbf{m} = (m_1, \ldots, m_r)$ with $m_1 \geqslant \cdots m_r \geqslant 0$, abbreviated as $\mathbf{m} \geqslant 0$, one defines a polynomial $\Delta_{\mathbf{m}}(x)$ on V by

$$(2) \qquad \Delta_{\mathbf{m}}(x) = \Delta_1^{m_1 - m_2}(x) \Delta_2^{m_2 - m_3}(x) \cdots \Delta_{r-1}^{m_{r-1} - m_r}(x) \Delta_r^{m_r}(x).$$

One observes that if $x \in \Omega$, then for any r-tuple of complex numbers $\mathbf{s} = (s_1, \ldots, s_r)$, replacing $\mathbf{m}$ by $\mathbf{s}$, (2) defines a function $\Delta_{\mathbf{s}}(x)$ on Ω.

For a real vector space E of dimension n with the inner product $(\cdot | \cdot)$ and a fixed orthonormal basis, $P(E)$ will denote the space of all complex-valued polynomials on E. For every polynomial $p \in P(E)$ we define a linear differential operator $p\left(\dfrac{\partial}{\partial x}\right)$ by expressing p in terms of the coordinates as $p(x_1, \ldots, x_n)$ and then formally replacing x_j by $\dfrac{\partial}{\partial x_j}$ $(1 \leqslant j \leqslant n)$.

For a polynomial $p(x, y)$ on $E \times E$, we can define a differential operator $p\left(x, \dfrac{\partial}{\partial x}\right)$ with polynomial coefficients in a similar way.

The Fischer inner product on $P(V)$ is defined by

$$(p, q)_F = \left(p\left(\frac{\partial}{\partial x}\right)\bar{q}\right)(0)$$

for $p, q \in P(V)$.

There is a natural representation π of the group $G(\Omega)$ on $P(V)$ defined by

$$(3) \qquad\qquad (\pi(g)p)(x) = p(g^{-1}.x)$$

for $g \in G(\Omega)$, $p \in P(V)$. The following is known, e.g. see [K].

THEOREM 1.1. *$P(V)$ is the orthogonal direct sum of the spaces $P_{\mathbf{m}}(V)$ ($\mathbf{m} \geqslant 0$) which are mutually inequivalent irreducible representation spaces of $G(\Omega)$. Moreover,*

$$P_{\mathbf{m}}(V) = \mathrm{span}\{\dot{\pi}(g)\Delta_{\mathbf{m}} \mid g \in G(\Omega)\}.$$

For each $\mathbf{m} \geqslant 0$, there is a unique L-invariant polynomial $\varphi_{\mathbf{m}}$ in $P_{\mathbf{m}}(V)$ with $\varphi_{\mathbf{m}}(e) = 1$ which is defined by

$$\varphi_{\mathbf{m}}(x) = \int_L \Delta_{\mathbf{m}}(l.x)dl.$$

For any r-tuple $\mathbf{s} = (s_1, \ldots, s_r) \in \mathbb{C}^r$, in the above formula, replacing $\mathbf{m}$ by $\mathbf{s}$ gives a spherical function $\varphi_{\mathbf{s}}$ on Ω, and moreover, every spherical function on Ω can be obtained in this way.

Under the Fischer inner product $(\cdot, \cdot)_F$, every $\mathcal{P}_{\mathbf{m}}(V)$ is a Hilbert space and has a reproducing kernel $K^{\mathbf{m}}(x, y) = K_y^{\mathbf{m}}(x)$, that is,

$$p(y) = \left(p, K_y^{\mathbf{m}}\right)_F \qquad \forall p \in \mathcal{P}_{\mathbf{m}}(V).$$

In fact, if $\left\{ \psi_{\mathbf{m}}^{(i)}(x), i = 1, \ldots, d_{\mathbf{m}} \right\}$ is an orthonormal basis of $\mathcal{P}_{\mathbf{m}}(V)$, where $d_{\mathbf{m}}$ is the dimension of $\mathcal{P}_{\mathbf{m}}(V)$, then

$$K^{\mathbf{m}}(x, y) = \sum_{i=1}^{d_{\mathbf{m}}} \psi_{\mathbf{m}}^{(i)}(x) \overline{\psi_{\mathbf{m}}^{(i)}(y)}.$$

Then one has, see [**K1**], that

(4) $$K^{\mathbf{m}}\left(g.x, {}^t g^{-1}.y\right) = K^{\mathbf{m}}(x, y)$$

for all $g \in G(\Omega)$. It is this property that makes $K^{\mathbf{m}}\left(x, \dfrac{\partial}{\partial x}\right)$ an element in $D(\Omega)$. In general, $p\left(\dfrac{\partial}{\partial x}\right)$ defined by a polynomial $p(x, y)$ is in $D(\Omega)$ if and only if p satisfies (4). For further detailed relations between polynomial invariants and invariant differential operators, we refer to [**HD, N, Y4**].

§1.2. The algebra $D(\Omega)$

In this section, we give some results about the algebra $D(\Omega)$ whose proofs are given in [**Y4**].

Considering $D(\Omega)$ as a real vector space, we have the following description:

PROPOSITION 1.2. $\left\{ K^{\mathbf{m}}\left(x, \dfrac{\partial}{\partial x}\right), \mathbf{m} \geqslant 0 \right\}$ is a basis of the vector space $D(\Omega)$.

REMARK 1. It follows from the proposition that every $D \in D(\Omega)$ can be extended to a differential operator on V with polynomial coefficients.

Among those $K^{\mathbf{m}}\left(x, \dfrac{\partial}{\partial x}\right)$, $\mathbf{m} \geqslant 0$, of particular interest to our study are $K^{1_j}\left(x, \dfrac{\partial}{\partial x}\right)$, $j = 1, \ldots, r$ where 1_j is the r-tuple of integers with 1 as its first j-th components and 0 the remaining components.

For $\lambda \in \mathbb{R}$, we define

$$D_\lambda = \Delta(x)^{1-\lambda} \Delta\left(\frac{\partial}{\partial x}\right) \circ \Delta(x)^\lambda,$$

then it is easy to verify that $D_\lambda \in D(\Omega)$.

The important identity

(5)
$$\Delta\left(\frac{\partial}{\partial x}\right)\Delta_{\mathbf{s}}(x) = \prod_{i=1}^{r}\left(s_i + \frac{r-i}{2}a\right)\Delta_{\mathbf{s}-1}(x)$$

is the starting point for our investigating $D(\Omega)$.

First, we relate $K^{1j}\left(x, \frac{\partial}{\partial x}\right)$ to D_λ by

THEOREM 1.3. *For any real number* λ,

(6)
$$D_\lambda = \sum_{j=0}^{r}\binom{r}{j}\prod_{l=1}^{r-j}\left(\lambda + \frac{l-1}{2}a\right)K_j$$

where $K_j = c_j K^{1j}\left(x, \frac{\partial}{\partial x}\right)$ *and* $c_j = \|\varphi_{1_j}\|_F^2$.

REMARK 2. The above expansion has also been obtained independently by J. Arazy.

Now we give some sets of generators of $D(\Omega)$.

THEOREM 1.4. *If* $\lambda_1, \ldots, \lambda_r$ *are distinct, then* $D_{\lambda_1}, \ldots, D_{\lambda_r}$ *are algebraically independent generators of* $D(\Omega)$.

The proof of this theorem uses (5) and the Harish-Chandra isomorphism, among other things.

Letting $\lambda = -\frac{i-1}{2}a$, $i = 1, \ldots, r$ in (6), we have r equations with a nonsingular coefficient matrix. Solving this system of equations, we obtain

THEOREM 1.5. K_i, $i = 1, \ldots, r$ *are algebraically independent generators of* $D(\Omega)$. *Moreover,*

$$K_j = \frac{j!}{r!}\left(\frac{2}{a}\right)^{r-j}\sum_{l=1}^{r-j+1}(-1)^l\binom{r-j}{l}D_{-\frac{l-1}{2}a}.$$

Motivated by Theorem 1.3, for any complex number λ, we define

$$D_\lambda = \sum_{j=0}^{r}\binom{r}{j}\prod_{l=1}^{r-j}\left(\lambda + \frac{l-1}{2}a\right)K_j.$$

The next two results are about the eigenvalues of those generators.

By (5), analytic continuation, Theorem 1.1 and Schur's lemma, we have

THEOREM 1.6. *For* $\mathbf{m} \geqslant 0$ *and any complex number* λ,

(7)
$$D_\lambda p = \prod_{i=1}^{r}\left(m_i + \lambda + \frac{r-i}{2}a\right)p, \qquad \forall p \in P_{\mathbf{m}}(V).$$

Next, we have

THEOREM 1.7. *For* $\mathbf{m} \geqslant 0$ *and* $j = 1, \ldots, r$,

$$(8) \qquad K_j p = \binom{r}{j}^{-1} \sum_{1 \leqslant i_1 < \cdots < i_j \leqslant r} \prod_{l=1}^{j} \left(m_{i_l} + \frac{j-l}{2}a \right) p, \qquad \forall p \in P_{\mathbf{m}}(V).$$

REMARK 3. Another set of generators is given in [N], but the corresponding eigenvalues are not available.

REMARK 4. In (7) and (8), replacing $\mathbf{m}$ by s and p by $\varphi_{\mathbf{s}}$, the same formulas still hold.

§1.3. Generalized hypergeometric functions associated with cones

In this section, we will give some results of generalized hypergeometric functions, which are related to our later discussions. For further results, see [GR, FK1, M, Y1, Y2].

For $\mathbf{s} = (s_1, \ldots, s_r) \in \mathbb{C}^r$, one defines

$$(9) \qquad (\mathbf{s})_{\mathbf{m}} = \prod_{i=1}^{r} \left(s_i - \frac{i-1}{2}a \right)_{m_i}$$

where $(c)_k = c(c+1)\cdots(c+k-1)$, $(c)_0 = 1$.

For $a_1, \ldots, a_p, b_1, \ldots, b_q \in \mathbb{C}$, such that $(b_j)_{\mathbf{m}} \neq 0$, for all $\mathbf{m}$, j, one defines the hypergeometric functions

$$(10) \qquad {}_pF_q(a_1, \ldots, a_p; b_1, \ldots, b_q; z) = \sum_{\mathbf{m}} \frac{(a_1)_{\mathbf{m}} \cdots (a_p)_{\mathbf{m}}}{(b_1)_{\mathbf{m}} \cdots (b_q)_{\mathbf{m}}} \frac{d_{\mathbf{m}}}{\left(\frac{n}{r}\right)_{\mathbf{m}}} \varphi_{\mathbf{m}}(z).$$

EXAMPLES. (i) ${}_0F_0(z) = e^{\langle z | e \rangle}$. (ii) ${}_1F_0(\mu; z) = \Delta(e - z)^{-\mu}$.

By the L-invariance of $\varphi_{\mathbf{m}}$, we see that ${}_pF_q(z) = {}_pF_q(t)$ if $z = l.t$ where $l \in L$, $t = \sum_{j=1}^{r} t_j c_j$. Thus we will also consider those hypergeometric functions as functions in r variables.

First, we have the following characterization of ${}_2F_1$

THEOREM 1.8. ${}_2F_1(a_1, a_2; b_1; x_1, \ldots, x_r)$ *is the unique solution of the system of the partial differential equations*

$$(11) \quad x_i(1 - x_i)\frac{\partial^2 F}{\partial x_i^2} + \left\{ b_1 - \frac{a}{2}(\kappa - 1) - \left[a_1 + a_2 + 1 - \frac{a}{2}(r-1) \right] x_i \right.$$

$$+ \frac{a}{2} \sum_{j=1, j \neq i}^{r} \frac{x_i(1 - x_i)}{x_i - x_j} \bigg\} \frac{\partial F}{\partial x_i} - \frac{a}{2} \sum_{j=1, j \neq i}^{r} \frac{x_j(1 - x_j)}{x_i - x_j} \frac{\partial F}{\partial x_j} = a_1 a_2 F$$

$i = 1, \ldots, r$, *subject to the conditions that*

(a) F *is a symmetric function of* $x_1, \ldots, x_r$ *and*
(b) F *is analytic at* $x_1 = \ldots = x_r = 0$ *and* $F(0) = 1$.

(11) is a generalization of the classical hypergeometric equation. When Ω is the cone of positive definite symmetric matrices, this result is due to Muirhead, see [**M**]. In general, this result was claimed in [**K2**], the complete proof is given by the author in [**Y1**]. Applying this theorem, we can easily show (ii) in the above examples and establish Kummer relations for generalized hypergeometric functions, see [**Y1**].

When $a_1 = a_2 = 0$, $\gamma = 1 + \frac{a}{2}(r-1)$, after a change of variable, the equations (11) become the radial parts, determined by M. Lassalle, of the Hua equations. The result that the constants are the only solutions is the key in Lassalle's proof that the Hua equations characterize the Poisson integrals on the Shilov boundary.

Next, we give the asymptotic behavior of $_{q+1}F_q$.

THEOREM 1.9. *Let* $\gamma = \sum_{i=1}^{p+1} a_i - \sum_{i=1}^{p} b_i$. *If for all* **m**

$$\frac{(a_1)_{\mathbf{m}} \cdots (a_{p+1})_{\mathbf{m}}}{(b_1)_{\mathbf{m}} \cdots (b_p)_{\mathbf{m}}} > 0.$$

For $-1 < y_i < 1$, $i = 1, \ldots, r$, *we have*

(i) *if* $\gamma > a/2(r-1)$, *then*

$$_{q+1}F_q(a_1, \ldots, a_{p+1}; b_1, \ldots, b_p; y_1, \ldots, y_r) \approx \prod_{i=1}^{r}(1 - y_i)^{-\gamma};$$

(ii) *if* $\gamma < -a/2(r-1)$, *then there exists a constant* C *such that*

$$_{q+1}F_q(a_1, \ldots, a_{p+1}; b_1, \ldots, b_p; y_1, \ldots, y_r) \leqslant C;$$

(iii) *if* $\gamma = a(-\frac{r-1}{2} + j - 1)$, $j = 1, \ldots, r$, *then, for* $y_1 = \cdots = y_r = t$, $-1 < t < 1$,

$$_{q+1}F_q(a_1, \ldots, a_{p+1}; b_1, \ldots, b_p; t, \ldots, t) \approx (1 - t)^{-(j-1)\frac{i}{2}a} \log \frac{1}{1-t};$$

(iv) *if* $a(-\frac{r-1}{2} + j - 1) < \gamma < (-\frac{r-1}{2} + j)a$, $j = 1, \ldots, r-1$, *then, for* $y_1 = \cdots = y = t$, $-1 < t < 1$,

$$_{q+1}F_q(a_1, \ldots, a_{p+1}; b_1, \ldots, b_p; t, \ldots, t) \approx (1 - t)^{-j[\gamma + a/2(r-j)]}.$$

If the rank is greater than one, the situation is quite complicated, and some new phenomena occur. When γ changes in the interval $[-\frac{r-1}{2}a, \frac{r-1}{2}a]$, we notice that the asymptotic behavior changes in a way different from that in the interval $(\frac{r-1}{2}a, \infty)$.

§2. Differential operators and spaces of holomorphic functions

In this part, we consider an irreducible bounded symmetric domain D in a complex vector space Z in the standard Harish-Chandra realization. Let G be the the identity component of $\text{Aut}(D)$ and K the isotropy subgroup of G at 0. For each complex number λ, using the expansion of D_λ obtained in part I, we introduce a differential operator $\mathcal{D}_\lambda$. For any distinct numbers $\lambda_1, \ldots, \lambda_r$, we also prove that $\mathcal{D}_{\lambda_1}, \ldots, \mathcal{D}_{\lambda_r}$ is a set of r algebraically independent generators of the algebra of holomorphic differential operators on Z that commute with the action of K. As in the case of a symmetric cone, one will see that polynomials in the irreducible subspaces of Schmid's decomposition are eigenfunctions of $\mathcal{D}_\lambda$, and, moreover, the eigenvalues can be calculated explicitly. We will use these differential operators to describe the dual and predual of the Bergman space $L^1(D) \cap H(D)$, characterize some Hilbert spaces of holomorphic functions occurring in representation theory and give integral formulas for invariant inner product on those Hilbert spaces.

We also consider spaces of holomorphic functions on some varieties in a bounded symmetric domain and obtain analogues of some classical results.

Finally, the generalized hypergeometric functions discussed in §1.3 will be related to some integrals on a bounded symmetric domain, then the asymptotic behavior of $_{q+1}F_q$ given there is applied to giving estimates of those integrals.

§2.1. Some background on bounded symmetric domains

Our notation follows that of [**FK2**].

Let $\mathcal{G}$ be the Lie algebra of G, and $\mathcal{K}$ the Lie algebra of K, then $\mathcal{G}$ is a simple real Lie algebra with Cartan decomposition

$$\mathcal{G} = \mathcal{K} + \mathcal{P}.$$

$\mathcal{G}^\mathbb{C}$ will be its complexification and $G^\mathbb{C}$ will be the adjoint group of $\mathcal{G}^\mathbb{C}$. A basis of root vectors $\{e_\alpha\}$ will be so chosen that $\tau e_\alpha = -e_{-\alpha}$, $[e_\alpha, e_{-\alpha}] = h_\alpha$, $[h_\alpha, e_{\pm\alpha}] = 2e_{\pm\alpha}$, where τ is the conjugation with respect to the real form $\mathcal{K} + i\mathcal{P}$.

Φ^+ will denote the set of positive non-compact roots, and setting

$$\mathcal{P}^\pm = \sum_{\alpha \in \Phi^+} \mathbb{C}e_{\pm\alpha},$$

one has

$$\mathcal{G}^\mathbb{C} = \mathcal{P}^- + \mathcal{K}^\mathbb{C} + \mathcal{P}^+.$$

Define a Hermitian inner product $(\cdot|\cdot)$ on $\mathcal{P}^+$ by $(z|w) = -B(z, \tau w)$ where B is the Killing form.

It is known that in the Harish-Chandra realization, D is a bounded symmetric domain in $\mathcal{P}^+$, and K acts on $\mathcal{P}^+$ by unitary transformations which coincide with the adjoint action.

Let $\gamma_1, \ldots, \gamma_r$ be the strongly orthogonal roots of Harish-Chandra with the ordering $\gamma_1 > \cdots > \gamma_r$. We simply write

$$e_j = e_{\gamma_j}, \quad (j = 1, \ldots, r), \qquad e = \sum_{j=1}^r e_j.$$

The Cayley transform is defined by $c = \exp i(\frac{\pi}{4})(e - \tau e)$. We write $^c\mathcal{G}$ for the Lie algebra of cGc^{-1} and $\mathcal{G}_T$ for the fixed point set of $\mathcal{G}$ under $\mathrm{Ad}(c^4)$.

Let $\mathcal{K}_T, \mathcal{P}_1, \mathcal{P}_1^+$ and $\mathcal{P}_1^-$ denote the intersections of $\mathcal{K}, \mathcal{P}, \mathcal{P}^+, \mathcal{P}^-$ with $\mathcal{G}_T^{\mathbb{C}}$ respectively, then one has the corresponding decompositions $\mathcal{G}_T = \mathcal{K}_T + \mathcal{P}_1$, $\mathcal{G}_T^{\mathbb{C}} = \mathcal{P}_1^- + \mathcal{K}_T^{\mathbb{C}} + \mathcal{P}_1^+$. $\mathrm{Ad}(c^2)$ is the Cartan involution of $\mathcal{K}_T$, the corresponding decomposition is $\mathcal{K}_T = \mathcal{L}_T + \mathcal{Q}_1$. Let $\mathcal{K}_T^* = \mathcal{L}_T + i\mathcal{Q}_1$ be its noncompact dual.

Now $\mathcal{N}_1^+ = {}^c\mathcal{G} \cap \mathcal{P}_1^+$ is a real form of $\mathcal{P}_1^+$. In particular, $\mathcal{N}_1^+$ has the structure of a real simple Euclidean Jordan algebra as described in [**KW**], e coincides with the identity element of the Jordan algebra $\mathcal{N}_1^+$ and $e_1, \ldots, e_r$ form Jordan frame.

K_T, K_T^* and G_T will denote the analytic subgroups in $G^{\mathbb{C}}$ corresponding to the Lie algebras $\mathcal{K}_T, \mathcal{K}_T^*$ and $\mathcal{G}_T$ respectively. Then $\Omega := K_T^*.e$ is the symmetric cone in $\mathcal{N}_1^+$, that is, the interior of the set of all squares in $\mathcal{N}_1^+$.

Let $\mathcal{H}^-$ be the real span of $h_{\gamma_1}, \ldots, h_{\gamma_r}$, then $i\mathcal{H}^-$ is a Cartan subalgebra of the pair $({}^c\mathcal{G}, {}^c\mathcal{K})$, and the $i\mathcal{H}^-$-roots of ${}^c\mathcal{G}$ are $\pm\frac{1}{2}(\gamma_j \pm \gamma_k)$, $\pm\gamma_j$, $\pm\frac{1}{2}\gamma_j$ $(1 \leqslant j, k \leqslant r)$ with respective multiplicities a, 1 and $2b$. See [**M**].

Let $\mathcal{P}^{+j/2}$ be the root space in $\mathcal{P}^+$ for $\frac{1}{2}\gamma_j$, and $\mathcal{P}_2^+ = \sum_j \mathcal{P}^{+j/2}$. Then $\mathcal{P}^+ = \mathcal{P}_1^+ + \mathcal{P}_2^+$, $n_1 := \dim \mathcal{P}_1^+ = \frac{r}{2}(r-1)a + r$ and $n := \dim \mathcal{P}_1^+ = \frac{r}{2}(r-1)a + r + rb$.

§2.2. Polynomials and their corresponding differential operators

Let U be a complex vector space of dimension n with a Hermitian inner product $(\cdot|\cdot)$ and coordinates $(z_1, \ldots, z_n)$, $P(U)$ the space of holomorphic polynomials on U, and $P(U \times \overline{U})$ the space of polynomials on $U \times U$ which are holomorphic in the first variable and antiholomorphic in the second variable.

We call D a holomorphic differential operator if in coordinates D can be expressed as

$$D = \sum_\alpha A_\alpha(z) \left(\frac{\partial}{\partial z}\right)^\alpha$$

where $\alpha = (\alpha_1, \ldots, \alpha_r)$ and $\left(\dfrac{\partial}{\partial z}\right)^\alpha = \left(\dfrac{\partial}{\partial z_1}\right)^{\alpha_1} \cdots \left(\dfrac{\partial}{\partial z_n}\right)^{\alpha_n}$.

For polynomials $p \in P(U)$ and $p(z, w) \in P(U \times \overline{U})$, we can define corresponding holomorphic differential operators $p\left(\dfrac{\partial}{\partial z}\right)$ and $p\left(z, \dfrac{\partial}{\partial z}\right)$ in manners similar to the real case.

Let u be an unitary operator on U and let it act on functions defined on U by $(u.f)(z) = f(u^{-1}.z)$. Then for $p \in P(U \times \overline{U})$, $p\left(z, \dfrac{\partial}{\partial z}\right)$ commutes with u if and only if

$$(1) \qquad p(u.z, u.w) = p(z, w), \qquad \forall z, w \in U.$$

The Fischer inner product on $P(U)$ is defined as follows

$$(2) \qquad (p, q)_{F,U} = \left(p\left(\frac{\partial}{\partial z}\right)\overline{q}\right)(0)$$

where $\overline{q}(z) = \overline{q(\overline{z})}$.

In the following, we will apply the above discussion to the complex vector spaces $\mathcal{P}_1^+$ and $\mathcal{P}^+$ without further mentioning.

Since $\mathcal{N}_1^+$ is a real form of $\mathcal{P}_1^+$, a holomorphic polynomial is determined by its restriction to $\mathcal{N}_1^+$, thus there is one-to-one correspondence between $P(\mathcal{P}_1^+)$ and $P(\mathcal{N}_1^+)$. For simplicity, we will use the same p to denote a polynomial in $P(\mathcal{N}_1^+)$ and its corresponding holomorphic polynomial in $P(\mathcal{P}_1^+)$.

Similarly, a complex-valued polynomial $p(x, y)$ on $\mathcal{N}_1^+ \times \mathcal{N}_1^+$ determines a unique polynomial $p(z, w)$ on $\mathcal{P}_1^+ \times \mathcal{P}_1^+$ which is holomorphic in z, antiholomorphic in w.

The Fischer inner products on $P_{\mathbf{m}}(\mathcal{N}_1^+)$ and $P_{\mathbf{m}}(\mathcal{P}_1^+)$ are denoted respectively by $(\cdot, \cdot)_{F, \mathcal{N}_1^+}$ and $(\cdot, \cdot,)_{F, \mathcal{P}_1^+}$.

A result analogous to Theorem 1.1 is the following which gives the Schmid decomposition, see [**FK2, S, U**].

THEOREM 2.1. *The space $P(\mathcal{P}^+)$ of holomorphic polynomials on $\mathcal{P}^+$ (resp. $\mathcal{P}_1^+$) decomposes into irreducible subspaces under $\mathrm{Ad}(K)$ (resp. $\mathrm{Ad}(K_T)$) as*

$$P(\mathcal{P}^+) = \bigoplus_{\mathbf{m} \geqslant 0} P_{\mathbf{m}}(\mathcal{P}^+)$$

and

$$P(\mathcal{P}_1^+) = \bigoplus_{\mathbf{m} \geqslant 0} P_{\mathbf{m}}(\mathcal{P}_1^+).$$

For each $\mathbf{m} \geqslant 0$, $\Delta_{\mathbf{m}} \in P_{\mathbf{m}}(\mathcal{P}_1^+)$, and its extension $\Delta_{\mathbf{m}}^E$ to $\mathcal{P}^+$ is in $P_{\mathbf{m}}(\mathcal{P}^+)$. For each $\mathbf{m} \geqslant 0$, restriction of polynomials maps $P_{\mathbf{m}}(\mathcal{P}^+)$ onto $P_{\mathbf{m}}(\mathcal{P}_1^+)$.

Let $K_1^{\mathbf{m}}(x, y)$, for $\mathbf{m} \geqslant 0$, be the reproducing kernel of $P_{\mathbf{m}}(\mathcal{N}_1^+)$ with respect to the Fischer inner product $(\cdot, \cdot)_{F, \mathcal{N}_1^+}$. An observation we make here is that $K_1^{\mathbf{m}}(z, w)$ is the reproducing kernel of $P_{\mathbf{m}}(\mathcal{P}_1^+)$ with respect to the Fischer inner product $(\cdot, \cdot)_{F, \mathcal{P}_1^+}$. This connects the differential operators in the real case with the differential operators to be defined in the following.

For each $\mathbf{m} \geqslant 0$, let $K^{\mathbf{m}}(z, w)$ be the reproducing kernel of $P_{\mathbf{m}}(\mathcal{P}^+)$, then each $K^{\mathbf{m}}$ satisfies (1).

When D is a classical domain, then K^{1j} can be written down explicitly.

For $j = 1, \ldots, r$, we define

$$\mathcal{K}_j = \frac{1}{c_{1_j}} K^{1j} \left(z, \frac{\partial}{\partial z} \right)$$

and

$$\mathcal{K}_j^T = \frac{1}{c_{1_j}} K_1^{1j} \left(z_1, \frac{\partial}{\partial z_1} \right).$$

Following Theorem 1.3, we define, for any complex number λ, holomorphic differential operators $\mathcal{D}_\lambda$ and $\mathcal{D}_\lambda^T$ respectively by

$$\mathcal{D}_\lambda = \sum_{j=0}^{r} \binom{r}{j} \prod_{l=1}^{r-j} \left(\lambda - \frac{r-l}{2} a \right) \mathcal{K}_j,$$

and

$$\mathcal{D}_\lambda^T = \sum_{j=0}^{r} \binom{r}{j} \prod_{l=1}^{r-j} \left(\lambda - \frac{r-l}{2} a \right) \mathcal{K}_j^T.$$

(We note that the parameter λ has been shifted by $-\frac{r-1}{2}a$.)

More generally, we define, for any positive integer k,

$$\mathcal{D}_\lambda^k = \mathcal{D}_\lambda \circ \mathcal{D}_{\lambda+1} \circ \cdots \mathcal{D}_{\lambda+k-1}.$$

It follows from Theorems 1.1, 2.1 and Schur's lemma that for each $\mathbf{n} \geqslant 0$, $K^{\mathbf{m}}\left(z, \frac{\partial}{\partial z} \right)$ (resp. $K_1^{\mathbf{m}}\left(z_1, \frac{\partial}{\partial z_1} \right)$ and $K_1^{\mathbf{m}}\left(x, \frac{\partial}{\partial x} \right)$) acts on $P_{\mathbf{n}}(\mathcal{P}^+)$ (resp. $P_{\mathbf{n}}(\mathcal{P}_1^+)$ and $P_{\mathbf{n}}(\mathcal{N}_1^+)$) as a scalar multiple of the identity, that is,

$$K^{\mathbf{m}}\left(z, \frac{\partial}{\partial z} \right)\bigg|_{P_{\mathbf{n}}(\mathcal{P}^+)} = \lambda_{\mathbf{n},\mathbf{m}}(\mathcal{P}^+) \ \mathrm{id}\ |_{P_{\mathbf{n}}(\mathcal{P}^+)},$$

$$K_1^{\mathbf{m}}\left(z_1, \frac{\partial}{\partial z_1} \right)\bigg|_{P_{\mathbf{n}}(\mathcal{P}_1^+)} = \lambda_{\mathbf{n},\mathbf{m}}(\mathcal{P}_1^+) \ \mathrm{id}\ |_{P_{\mathbf{n}}(\mathcal{P}_1^+)},$$

and

$$K^{\mathbf{m}}\left(x, \frac{\partial}{\partial x} \right)\bigg|_{P_{\mathbf{n}}(\mathcal{N}_1^+)} = \lambda_{\mathbf{n},\mathbf{m}}(\mathcal{N}_1^+) \ \mathrm{id}\ |_{P_{\mathbf{n}}(\mathcal{N}_1^+)},$$

for some constants $\lambda_{\mathbf{n},\mathbf{m}}(\mathcal{P}^+)$, $\lambda_{\mathbf{n},\mathbf{m}}(\mathcal{N}_1^+)$ and $\lambda_{\mathbf{n},\mathbf{m}}(\mathcal{P}_1^+)$.

Now we see that $\mathcal{K}_j$ and $\mathcal{D}_\lambda$ (resp. $\mathcal{K}_j^T$ and $\mathcal{D}_\lambda^T$) are diagonal on the polynomial space $P(\mathcal{P}^+)$ (resp. $P(\mathcal{P}_1^+)$) corresponding to the Schmid decomposition. The main purpose is to calculate the eigenvalues of $\mathcal{D}_\lambda$ and $\mathcal{K}_j$ on the corresponding irreducible spaces. Roughly speaking, the actions of $\mathcal{D}_\lambda^T$ and $\mathcal{K}_j^T$ on $P_{\mathbf{m}}(\mathcal{P}_1^+)$ are almost the same as those of D_λ and K_j on $P_{\mathbf{m}}(\mathcal{N}_1^+)$, then the eigenvalues of $\mathcal{D}_\lambda^T$ and and $\mathcal{K}_j^T$ can be immediately obtained from the results in §1.2. Thus the key step is to find the relation between the eigenvalues of $K^{\mathbf{m}}\left(z, \frac{\partial}{\partial z} \right)$ and $K_1^{\mathbf{m}}\left(z_1, \frac{\partial}{\partial z_1} \right)$, which is given by the following result

THEOREM 2.2. *For all* $\mathbf{m} \geqslant 0$, $\mathbf{n} \geqslant 0$,

$$(3) \qquad \lambda_{\mathbf{n},\mathbf{m}}(\mathcal{P}^+) = \lambda_{\mathbf{n},\mathbf{m}}(\mathcal{P}_1^+) = \lambda_{\mathbf{n},\mathbf{m}}(\mathcal{N}_1^+)$$

As a consequence of Theorems 1.6, 1.7 and 2.2, we have

THEOREM 2.3. *For* $\mathbf{m} \geqslant 0$, *and any* $p \in P_{\mathbf{m}}(\mathcal{P}^+)$,

(i)

$$(4) \qquad \mathcal{D}_\lambda^{(k)} p = \mu_{\mathbf{m}}^{(k)}(\lambda) p$$

$$where\ \mu_{\mathbf{m}}^{(k)}(\lambda) = \prod_{i=1}^{r} \prod_{j=0}^{k-1} \left(\lambda + m_i + j - \frac{i-1}{2} a \right);$$

(ii)

$$\mathcal{K}_j p = \frac{j!}{r!} \left(\frac{2}{a}\right)^{r-j} \sum_{l=0}^{r-j} (-1)^l \binom{r-j}{l} \prod_{l=1}^{r} \left(m_i + \frac{r-i-l}{2} a\right) p$$

(5)

$$= \binom{r}{j}^{-1} \sum_{1 \leqslant i_1 < \cdots < i_j \leqslant r} \prod_{l=1}^{j} \left(m_{i_l} + \frac{j-l}{2} a\right) p.$$

As for the algebra $\mathcal{D}^K$ of holomorphic differential operators that commute with the action of K, we have the following results similar to those of $D(\Omega)$.

PROPOSITION 2.4. $\left\{ K^{\mathbf{m}}\left(z, \dfrac{\partial}{\partial z}\right), \mathbf{m} \geqslant 0 \right\}$ is a basis of $\mathcal{D}^K$.

THEOREM 2.5.

(i) $\mathcal{D}^K$ is a commutative algebra;

(ii) $\left\{ K^{1_1}\left(z, \dfrac{\partial}{\partial z}\right), \ldots, K^{1_r}\left(z, \dfrac{\partial}{\partial z}\right) \right\}$ is a set of algebraically independent generators of $\mathcal{D}^K$.

THEOREM 2.6. For any distinct numbers $\lambda_1, \ldots, \lambda_r$, $\mathcal{D}_{\lambda_1}, \ldots, \mathcal{D}_{\lambda_r}$ is a set of algebraically independent generators of $\mathcal{D}^K$.

§2.3. Hilbert spaces of holomorphic functions

We denote by $h(z)$ the K-invariant polynomial on $\mathcal{P}^+$ whose restriction to $\{\sum_{i=1}^{r} a_i e_i \mid a_i \in \mathbb{R}, i = 1, \ldots, r\}$ is given by

$$h\left(\sum_{i=1}^{r} a_i c_i\right) = \prod_{i=1}^{r} (1 - a_i^2).$$

Let

$$h(z, w) = \exp \sum_{j=1}^{n} z_j \frac{\partial}{\partial z_j} \exp \sum_{j=1}^{n} \overline{w}_j \frac{\partial}{\partial \overline{z}_j} h(z),$$

then

(6) $$h(z, w)^{-p} = K(z, w),$$

where $p = (r-1)a + b + 2$ and $K(z, w)$ is the Bergman kernel of D. See [**FK**].

One defines the Hilbert space H_λ, parametrized by a real number $\lambda > p - 1$, of holomorphic functions f on D such that

(7) $$\|f\|_\lambda^2 = c_\lambda \int_D |f(z)|^2 h(z)^{\lambda - p} dz < \infty$$

where c_λ is some constant depending on λ, and the inner product is defined by (7). Then H_λ has $K_\lambda(z, w) = h(z, w)^{-\lambda}$ as its reproducing kernel. For an element $g \in G$, one can define a linear transformation $U_\lambda(g)$ on H_λ by

(8) $$(U_\lambda(g)f)(z) = f(g^{-1}z) J_{g-1}(z)^{\lambda/p}, f \in H_\lambda$$

here J_g is the complex Jacobian determinant of g and we use the principal branch of the power functions. Then (8) defines a unitary representation U_λ of $\tilde{G}$, the universal covering group of G, on H_λ, which is so-called holomorphic discrete series.

We notice that when $\lambda \leqslant p - 1$, the integral in (7) diverges for nonzero f. It is desirable to seek some other Hilbert spaces of holomorphic functions on D, parametrized by $\lambda \leqslant p - 1$, such that first they will be "analytic continuation" of those $H_\lambda, \lambda > p - 1$, and secondly, (8) will still define unitary representations of $\tilde{G}$ on them, thus new representations will be obtained. These have been done by N. Wallach in an algebraic approach [**W**], and by J. Faraut and A. Korányi in an analytic approach [**FK2**], and also by some others.

We will consider those Hilbert spaces following [**FK2**].

For $\lambda \in \mathbb{C}$, we denote by $\mathcal{P}^{(\lambda)}$ the set $P(\mathcal{P}^+)$ equipped with the structure of a Harish-Chandra module given in [**FK**]. For $\mathbf{m} \geqslant 0$, let $q(\lambda, \mathbf{m})$ be the multiplicity of λ as a zero of the polynomial $\lambda' \rightarrow (\lambda')_{\mathbf{m}}$. Set $q(\lambda) = \sup_{\mathbf{m} \geqslant 0} q(\lambda, \mathbf{m})$. Clearly, $q(\lambda) \leqslant r$. For $j = 0, 1, \ldots, q(\lambda)$, let

$$M_j^{(\lambda)} = \left\{ f \in \mathcal{P}^{(\lambda)} \,\middle|\, f = \sum_{\mathbf{m} \geqslant 0, q(\lambda, \mathbf{m}) \leqslant j} f_{\mathbf{m}}, f_{\mathbf{m}} \in P_{\mathbf{m}}(\mathcal{P}^+) \right\}.$$

According to Theorem 5.3 in [**FK**], $q(\lambda) > 0$ if and only if $\lambda - \frac{r-1}{2}a$ or $\lambda - \frac{r-2}{2}a$ is a nonpositive integer, and

$$M_0^{(\lambda)} \subset M_1^{(\lambda)} \subset \cdots \subset M_{q(\lambda)}^{(\lambda)} = \mathcal{P}^{(\lambda)}$$

is a composition series of $\mathcal{P}^{(\lambda)}$. Moreover, for every integer $0 \leqslant j \leqslant (\lambda)$, $M_j^{(\lambda)}/M_{j-1}^{(\lambda)}$ $(M_{-1} := 0)$ has a U_λ-invariant Hermitian form given by

$$(9) \qquad (f, g)_{\lambda, j} = \lim_{\lambda' \to \lambda} \frac{(\lambda' - \lambda)^j}{(\lambda)_{\mathbf{m}}} (f, g)_F$$

for $f, g \in P_{\mathbf{m}}(\mathcal{P})$.

The Hermitian form $(\cdot, \cdot)_{\lambda, 0}$ on M_0 is definite if and only if $\lambda > \frac{r-1}{2}a$ or $\lambda = j\frac{a}{2}$ with an integer $0 \leqslant j \leqslant r - 1$. For $j \geqslant 1$, $(\cdot, \cdot)_{\lambda, j}$ on $M_j^{(\lambda)}/M_{j-1}^{(\lambda)}$ is definite if and only if $j = q(\lambda)$ and $\frac{r-1}{2}a - \lambda$ is an integer. In either case, $j = 0$ or $q(\lambda)$, $M_j^{(\lambda)}/M_{j-1}^{(\lambda)}$ is said to be unitarizable. Of particular interest is the case when the quotient is unitarizable, that is, the corresponding Hermitian form is an inner product. In this case, one will obtain a corresponding Hilbert space of holomorphic functions on which (8) will define a unitary representation $\tilde{G}$.

Our purpose in this section is to express the invariant Hermitian form (9) in terms of integrals on D and describe the corresponding Hilbert space.

First we study those unitarizable M_0^λ.

We observe that when $\lambda > \frac{r-1}{2}a$, M_0^λ is the polynomial space $P(\mathcal{P}^+)$ itself. For $\lambda = \frac{i-1}{2}a, 1 \leqslant j \leqslant r$, then we will characterize $M_0^{(\frac{r-1}{2}a)}$ as the kernel of our differential operator $\mathcal{K}_j$.

If $\lambda = \frac{i-1}{2}a$, $1 \leqslant j \leqslant r$, then for $\mathbf{m} = (m_1, \ldots, m_{j-1}, 0, \ldots, 0)$, $(\lambda)_\mathbf{m} > 0$, and for all other $\mathbf{m}$, $(\lambda)_\mathbf{m} = 0$. Hence we have

$$M_0^{\left(\frac{i-1}{2}a\right)} = \bigoplus_{\mathbf{m} \geqslant 0, m_j = \cdots = m_r = 0} P_\mathbf{m}(\mathcal{P}^+).$$

It is shown in [U] that when D is of tube type, $M_0^{\left(\frac{r-1}{2}a\right)}$ is the space of harmonic polynomials, in the sense of

(10) $$\Delta\left(\frac{\partial}{\partial z}\right) p(z) = 0.$$

We note that when D is of tube type for $p \in P(\mathcal{P}^+)$, (10) is equivalent to $\mathcal{K}_r p = 0$. Now we generalize this result as follows

THEOREM 2.7. *For* $\lambda = \frac{i-1}{2}a$, $j = 1, \ldots, r$, *we have*

(11) $$M_0^{\left(\frac{i-1}{2}a\right)} = \{p \in P(\mathcal{P}^+) \mid \mathcal{K}_j p = 0.\}$$

The proof follows from Theorem 2.3.

Secondly, we consider the completion H_λ of $M_0^{(\lambda)}$ with respect to the inner product

$$(f, g)_{\lambda,0} = \sum_\mathbf{m} \frac{(f_\mathbf{m}, g_\mathbf{m})_F}{(\lambda)_\mathbf{m}}, \qquad f, g \in M_0^{(\lambda)},$$

for $\lambda > \frac{r-1}{2}a$ or $\lambda = \frac{i}{2}a$, $0 \leqslant j \leqslant r - 1$.

By Theorem 5.2.7 [K1], every $f \in H(D)$ can be expanded as

$$f(z) = \sum_{\mathbf{m} \geqslant 0} f_\mathbf{m}$$

where $f_\mathbf{m} \in \mathcal{P}_\mathbf{m}$, and the series converges uniformly and absolutely on compact subsets of D.

For $f, g \in H(D)$, let $f = \sum_{\mathbf{m} \geqslant 0} f_\mathbf{m}$, $g = \sum_{\mathbf{m} \geqslant 0} g_\mathbf{m}$ be the expansion as above, it can be readily seen that

(i) for $\lambda > \frac{r-1}{2}a$

$$H_\lambda = \left\{f \in H(D) \,\Big|\, \sum_\mathbf{m} \frac{(f_\mathbf{m}, f_\mathbf{m})_F}{(\lambda)_\mathbf{m}} < \infty\right\}$$

with the inner product

(12) $$(f, g)_\lambda = \sum_\mathbf{m} \frac{(f_\mathbf{m}, g_\mathbf{m})_F}{(\lambda)_\mathbf{m}};$$

(ii) for $\lambda = \frac{i}{2}a$, $0 \leqslant j \leqslant r - 1$,

$$H_\lambda = \left\{f \in H_j(D) \,\Big|\, \sum_\mathbf{m} \frac{(f_\mathbf{m}, f_\mathbf{m})_F}{(\lambda)_\mathbf{m}} < \infty\right\}$$

with the inner product

(13)
$$(f,g)_\lambda = \sum_{\mathbf{m}} \frac{(f_{\mathbf{m}}, g_{\mathbf{m}})_F}{(\lambda)_{\mathbf{m}}},$$

where $H_j(D)$ consists of holomorphic functions $f = \sum_{\mathbf{m} \geq 0} f_{\mathbf{m}}$ which have terms $f_{\mathbf{m}} \neq 0$ only for those $\mathbf{m}$ with $m_{j+1} = \cdots = m_r = 0$ or equivalently, $\mathcal{K}_j f = 0$.

REMARK 1. When $\lambda > p - 1$, the new definition of H_λ coincides with the previous one.

For $\lambda > \frac{r-1}{2}a$ or $\lambda = \frac{i}{2}a$, $0 \leq j \leq r-1$, $K_\lambda(z,w) := h(z,w)^{-\lambda}$ is the reproducing kernel on H_λ, and the closure of the linear span $\{K_\lambda(\cdot, w), w \in D\}$ is H_λ. Now by the identities $(K_\lambda(\cdot, w), K_\lambda(\cdot, z))_\lambda = K_\lambda(z, w)$ and $J_g(z)K(g.z, g.w) \cdot \overline{J_g(w)} = K(z, w)$, $z, w \in D$, or by Theorem 5.3 in [**FK**], we conclude that $(\cdot, \cdot)_\lambda$ is invariant under the action U_λ.

Now we give the following

THEOREM 2.8. *For $\lambda > \frac{r-1}{2}a$, $k \in \mathbb{Z}$ with $\lambda + k > p - 1$,*

(14)
$$H_\lambda = \left\{ f \in H(D) \,\middle|\, \int_D (\mathcal{D}_\lambda^k f)(z)\overline{g(z)}h(z,z)^{\lambda+k-p}dV(z) < \infty \right\}$$

and

(15)
$$(f,g)_\lambda = c_{\lambda,k} \int_D (\mathcal{D}_\lambda^k f)(z)\overline{g(z)}h(z,z)^{\lambda+k-p}dV(z)$$

*where $c_{\lambda,k}$ is a constant, see [**Y4**].*

REMARK 2. When f, g are not in $H(\overline{D})$, the integrals in (28) and (29) are understood as $\lim_{r \to 1} \int_D (\mathcal{D}_\lambda^k f)(rz)\overline{f(rz)}h(z,z)^{\lambda+k-p}dV(z)$.

For $\lambda > \frac{r-1}{2}a$ or $\lambda = \frac{i}{2}a$, $0 \leq j \leq r-1$, $k \in \mathbb{Z}$ with $\lambda + 2k > p - 1$, we define an inner product $\|\cdot\|_{\lambda,k}$ on H_λ by

$$\|f\|^2_{\lambda,k} = \int_D (\mathcal{D}_{n_1/r}^k f)(z)\overline{\mathcal{D}_{n_1/r}^k f)(z)}h(z,z)^{\lambda+2k-p}dV(z).$$

Applying the Stirling's formula, we can find two positive constants C_1 and C_2 such that

(16)
$$C_1\|f\|^2_{\lambda,k} \leq (f,f)_\lambda \leq C_2\|f\|^2_{\lambda,k}.$$

This yields

THEOREM 2.9. *For $k \in \mathbb{Z}$ with $\lambda + 2k > p - 1$, we have*

(i) *if $\lambda = \frac{i-1}{2}a$, $1 \leq j \leq r$, then*

$$H_\lambda = \{f \in H(D) \mid \mathcal{K}_j f = 0, \|f\|_{\lambda,k} < \infty\};$$

(ii) *if $\lambda > \frac{r-1}{2}a$, then*

$$H_\lambda = \{f \in H(D) \mid \|f\|_{\lambda,k} < \infty\}.$$

Finally, we consider the case that $M_{q(\lambda)}^{(\lambda)}/M_{q(\lambda)-1}^{(\lambda)}$ is unitarizable.

For $\lambda = n_1/r - s$, $s \geqslant 1$, we denote by $\tilde{H}_\lambda$ the completion of $M_{q(\lambda)}^{(\lambda)}/M_{q(\lambda)-1}^{(\lambda)}$ with respect to $(\cdot, \cdot)_{\lambda, q(\lambda)}$.

We have

THEOREM 2.10. $\tilde{H}_\lambda$ is identified with the space of holomorphic functions for which

$$\int_D (\mathcal{D}_\lambda^k f)(z)\overline{f(z)} h(z, z)^{\lambda+k-p} dV(z) < \infty$$

with the invariant inner product

$$(f, g)_{\lambda, q(\lambda)} = c \int_D (\mathcal{D}_\lambda^k f)(z)\overline{g(z)} h(z, z)^{\lambda+k-p} dV(z),$$

where $k \in \mathbb{Z}$ with $\lambda + k > p - 1$ and c is a constant depending on λ and k. When f, g are not in $H(\overline{D})$, the integral has the same meaning as in Remark 2.

When $r = 1$, $s = 1$, then D is the unit disc and $\lambda = 0$, the Hermitian inner product in (9) can be given by the integral

(17)
$$\int_D f'(z)\overline{g'(z)} dV(z),$$

hence $\tilde{H}_\lambda$ is just the classical Dirichlet space. It is interesting to express (9) in the general case by an integral analogous to (17). This is done in the next result, thus we will call $\tilde{H}_\lambda$ Generalized Dirichlet Space.

THEOREM 2.11. If $\lambda = n_1/r - s \leqslant 0$ and $s \geqslant n/r - 1$, let

(18)
$$(f, g)_{\lambda, q(\lambda)} = c \int_D (\mathcal{D}_{\lambda+b}^s f)(z)\overline{(\mathcal{D}_\lambda^s g)(z)} \frac{h(z, z)^{s-n/r}}{|z|^{2s}} dV(z)$$

where c depends on λ, s. Then $\tilde{H}_\lambda$ is identified with the space of holomorphic functions on D for which $(f, f)_{\lambda, q(\lambda)}$ is finite and the invariant inner product is given by (18). In particular, when D is of tube type, then (18) becomes

(19) $$(f, g)_{\lambda, q(\lambda)} = \int_D \left(\Delta\left(\frac{\partial}{\partial z}\right)^s f\right)(z)\overline{\left(\Delta\left(\frac{\partial}{\partial z}\right)^s g\right)(z)} h(z, z)^{s-n/r} dV(z).$$

REMARK 3. When D is of tube type and $n/r - s$ is an integer, (19) is due to J. Arazy.

For other integral formulas of the invariant inner products, we refer to [A1, A2]. For further topics related to representation theory, we refer to [DES, RV, W]. Hankel and Toeplitz operators on the Hilber space H_p have been extensively studied, we refer the reader to [BCZ, Zh, Z3].

§2.4. The dual and predual of the Bergman spaces

For $q \geqslant 1$, let $L_a^q(D) = L^q(D) \cap H(D)$ be the Bergman space on the bounded symmetric domain D. When $q > 1$, using weighted Bergman projection operator, one can easily show that the dual of $L^q(D)$ is $L^{q'}(D)$ where $\frac{1}{q} + \frac{1}{q'} = 1$, but, when $q = 1$, the situation becomes quite subtle. However, we can describe, in a way similar to the one variable case, the predual and dual of the Bergman space $L_a^1(D)$ in terms of some differential operators introduced in §2.2.

For characterizations of the dual and the predual of the Bergman space $L_a^1(D)$ in terms of Hankel operators, see [**Z1**].

First, we give the following result which is interesting in its own right.

THEOREM 2.12. *For any complex number λ and any positive integer k,*

$$\mathcal{D}_{\lambda,z}^k h(z,w)^{-\lambda} = c_{\lambda,k} h(z,w)^{-(\lambda+k)}$$

where $c_{\lambda,k} = \prod_{j=1}^k \prod_{i=1}^r (\lambda + j - 1 - \frac{i-1}{2})$.

This follows from Theorem 2.3 and the expansion

$$h(z,w)^{-\lambda} = \sum_{\mathbf{m}} (\lambda)_{\mathbf{m}} K^{\mathbf{m}}(z,w)$$

which is due to Faraut and Korányi, [**FK2**].

Next, we introduce Bloch-type spaces of holomorphic functions on D. Writing $\mathcal{D}^s$ for $\mathcal{D}_p^s$, we define

$$\tilde{\mathcal{B}}^s(D) = \{f \in L_a^2(D) \mid \sup_{z \in D} h(z,z)^s |(\mathcal{D}^s f)(z)| < \infty\}$$

and

$$\tilde{\mathcal{B}}_0^s(D) = \{f \in L^2(D) \mid \lim_{z \to \partial D} h(z,z)^s |(\mathcal{D}^s f)(z)| = 0\},$$

where ∂D is the topological boundary of D.

Then $\tilde{\mathcal{B}}^s(D)$ and $\tilde{\mathcal{B}}_0^s(D)$ become Banach spaces with the norm

$$\|f\|_* = \sup_{z \in D} h(z,z)^s |(\mathcal{D}^s f)(z)|.$$

Let P be the Bergman projection, $C(\bar{D})$ the space of continuous functions on $\bar{D}$ and $C_0(D)$ the subspace of $C(\bar{D})$ consisting of functions which vanish on ∂D. As in the classical case, we have

THEOREM 2.13. *For $s > \frac{r-1}{2} a$,*

$$P : L^\infty(D) \to \tilde{\mathcal{B}}^s(D),$$
$$P : C(\bar{D}) \to \tilde{\mathcal{B}}_0^s(D),$$
$$P : C_0(D) \to \tilde{\mathcal{B}}_0^s(D),$$

are bounded and onto. Therefore, for all $s > \frac{r-1}{2}a$, the $\tilde{\mathcal{B}}^s(D)$ are the same and $\tilde{\mathcal{B}}_0^s(D)$ are the same.

For a Banach space X, we write X^* for its dual.

Next result gives the dual and predual of the Bergman space $L_a^1(D)$.

THEOREM 2.14. *For $s > \frac{r-1}{2}a$, $L_a^1(D)^* = \tilde{\mathcal{B}}^s(D)$ and $\tilde{\mathcal{B}}_0^s(D)^* = L_a^1(D)$.*

For the proofs, see [**Y3, Y4**].

REMARK. Since the differential operators $\mathcal{D}_\lambda^k$ have the same actions on holomorphic functions as the integral operators studied in [**Z2**], one can also use $\mathcal{D}_\lambda^k$ to characterize the holomorphic Besov spaces introduced in [**Z2**].

§2.5. Estimates of some integrals

For $\gamma \in \mathbb{R}$, $\lambda > p - 1$ and $z \in D$, one defines

$$(20) \qquad I_\gamma(z) = \int_S |h(z,u)|^{-\left(\frac{n}{r}+\gamma\right)} du,$$

$$(21) \qquad J_{\gamma,\lambda}(z) = \int_D |h(z,u)|^{-(\lambda+\gamma)} h(w,w)^{\lambda-p} dw.$$

The estimates of these two integrals are very important to the study of the boundedness of the Bergman projection and some integral operators in which the function h is involved.

The following lemma gives the relation between these integrals and generalized hypergeometric functions, see [**K2**].

LEMMA. *If $z = k.t$ where $t = \sum_{i=1}^r t_i e_i$, then*

$$I_\gamma(z) = {}_2F_1\left(\frac{\frac{n}{r}+\gamma}{2}, \frac{\frac{n}{r}+\gamma}{2}; \frac{\frac{n}{r}+\gamma}{2}; t^2\right),$$

$$J_{\gamma,\lambda}(z) = {}_2F_1\left(\frac{\lambda+\gamma}{2}, \frac{\lambda+\gamma}{2}; \lambda; t^2\right).$$

Then the the asymptotic behavior of (20) and (21), as z approaches the boundary of D, can be seen immediately from Theorem 1.9. The estimate for $|\gamma| > \frac{r-1}{2}a$ is due to Faraut and Korányi, and in general, is due to the author. Using our results, we can generalize the results in [**M, MS**] to all bounded symmetric domains in a unified way. For further details, we refer to [**FK, M, MS**].

§2.6. Function spaces on varieties in a bounded symmetric domain

Let

$$S_j = \left\{ z \in \mathcal{P}^+ \;\middle|\; z = k.\sum_{i=1}^j e_i, k \in K \right\},$$

$$\bar{S}_j^{\mathbb{C}} = \left\{ z \in \mathcal{P}^+ \;\middle|\; z = k.\sum_{i=1}^j t_i e_i, t_i \in \mathbb{R}, k \in K \right\}.$$

Then it is shown by Upmeier that $\bar{S}_j^{\mathbb{C}}$ is an algebraic variety and S_j is the Shilov boundary of the bounded domain $V_j := D \cap \bar{S}_j^{\mathbb{C}}$ in $\bar{S}_j^{\mathbb{C}}$.

For $1 \leqslant j \leqslant r$, endow S_j with the K-invariant probability measure σ_j, and let $H^2(S_j)$ be the closed subspace of $L^2(S_j)$ spanned by the polynomials in $M^{(\frac{1}{2}a)}$. These Hilbert spaces have occurred in the study of Toeplitz operators on bounded symmetric domains by Upmeier.

Here we will investigate some Hilbert spaces and Banach spaces of holomorphic functions on V_j including $H^2(S_j)$ from function - theoretic point of view.

Now let $H(V_j)$ be the space of holomorphic functions on V_j. Recall that $f \in H(V_j)$ means that for every point $z \in V_j$ there exists an open neighborhood U_z in D and a holomorphic function f_z on U_z such that $f_z|_{U_z \cap V_j} = f|_{U_z \cap V_j}$. For an arbitrary analytic variety $V \in D$ and a holomorphic function f on V, in general, one can not find a global holomorphic function F on D with $F|_V = f$. However, for these special V_j, we have

THEOREM 2.15. *For any $f \in H(V_j)$, there exists a unique F in $H_j(D)$ such that $F|_{V_j} = f$.*

We note that S_r is the Shilov boundary of the domain D. In this case, it is known that $H^2(S_r)$ can be identified with the space of boundary values of functions in the ordinary Hardy space on D. Naturally, one may ask if this result is still true for $H^2(S_j)$ with $j = 1, \ldots, r-1$. It turns out that it is indeed the case. However, we will give our results in a more general setting.

Let $\varphi : (-\infty, \infty) \to [0, \infty)$ be a nondecreasing convex function not identically zero, and let $H_\varphi(V_j)$ be defined as follows

$$(22) \qquad H_\varphi(V_j) = \left\{ f \in H(V_j) \mid \sup_{0 < r < 1} \int_{S_j} \varphi(\log |f(r\zeta)|) d\sigma_j(\zeta) < \infty \right\}.$$

(i) If $\varphi(x) = \max(0, x)$, then let $N(V_j) := H_\varphi(V_j)$.
(ii) If $\varphi(x) = e^{px}$ with $p > 0$, then let $H^p(V_j) := H_\varphi(V_j)$ and $\|f\|_p^p = \sup_{0<r<1} \int_{S_j} |f(r\zeta)|^p d\sigma_j(\zeta)$.

THEOREM 2.16.

(i) *Suppose that $f \in N(V_j)$, then, for almost all $\zeta \in S_j$, $\lim_{r \to 1} f(r\zeta)$ exists, denoted $f^*(\zeta)$;*
(ii) *Suppose that $f \in H^p(V_j)$ ($p > 0$), if f^* is identically zero on a subset of S_j of positive measure, then f is identically zero on V_j.*

THEOREM 2.17. *If $f \in H^p(V_j)$, then*

(i) *$f^* \in L^p(S_j)$;*
(ii) *$\lim_{r \to 1} \int_{S_j} |f^*(\zeta) - f(r\zeta)| d\sigma_j(\zeta)$ and $\|f^*\|_{L^p(S_j)} = \|f\|_p$.*

For $j = 1, \ldots, r$, define

$$C_j(z, \zeta) = \sum_{\mathbf{m} \geq 0} \left(\frac{n}{r}\right)_{\mathbf{m}} \frac{\left(\frac{r}{2}a\right)_{\mathbf{m}}}{\left(\frac{j}{2}a\right)_{\mathbf{m}}} K^{\mathbf{m}}(z, \zeta).$$

THEOREM 2.18. *If $f \in H^p(S_j), p \geq 1$, then*

(23)
$$f(z) = \int_{S_j} C_j(z, \zeta) f^*(\zeta) d\sigma_j(\zeta).$$

COROLLARY. *$C_j(z, \zeta)$ is the reproducing kernel of the Hilbert space $H^2(S_j)$. Moreover, for $z = k.t$ with $t = \sum_{i=1}^j t_j e_j$,*

$$C_j(z, z) = {}_2F_1\left(\frac{n}{r}, \frac{r}{2}a; \frac{j}{2}a; t^2\right).$$

For each $j = 1, \ldots, r$, we can also consider Bergman spaces on the variety V_j. First we have

THEOREM 2.19. *Let v_j denote the measure inherited from the Lebesgue measure in $\mathcal{P}^+$, then*

$$\int_{V_j} f(z) dv_j(z) = c \int_K \int_0^1 \cdots \int_0^1 f\left(k. \sum_{i=1}^j t_i e_i\right) \prod_{i=1}^j t_i^{2b+2a(r-j)+1}$$

$$\times \prod_{1 \leq i < l \leq j} |t_i^2 - t_l^2|^a dt_1 \cdots dt_j.$$

More generally, we let $d\mu_{\alpha,\beta}$ denote the K-invariant measure on V_j with total mass 1 such that

$$\int_{V_j} f(z) d_j^{(\alpha,\beta)} \mu(z) = c \int_K \int_0^1 \cdots \int_0^1 f\left(k. \sum_{i=1}^j t_i e_i\right) \prod_{i=1}^j (1 - t_i^2)^\alpha$$

$$\times \prod_{i=1}^j t_i^{2\beta+1} \prod_{1 \leq i < l \leq j} |t_i^2 - t_l^2|^a dt_1 \cdots dt_j.$$

Now we define the weighted Bergman space $L_a^p\left(V_j, d\mu_j^{(\alpha,\beta)}\right)$ on V_j for $\alpha > -1$, $\beta > -1$, by

$$L_a^p\left(V_j, d\mu_j^{(\alpha,\beta)}\right) = \left\{f \in H(V_j) \,\Big|\, \int_{V_j} |f(z)|^p d\mu_j^{(\alpha,\beta)}(z) < \infty\right\},$$

and let

$$K_j^{(\alpha,\beta)}(z, w) = \sum_{\mathbf{m} \geq 0} \left(\frac{n}{r}\right)_{\mathbf{m}} \frac{(\alpha + \beta + (j-1)a + 2)_{\mathbf{m}} \left(\frac{r}{2}a\right)_{\mathbf{m}}}{\left(\beta + \frac{i-1}{2}a + 1\right)_{\mathbf{m}} \left(\frac{i}{2}a\right)_{\mathbf{m}}} K^{\mathbf{m}}(z, w).$$

Then for $z = k.t$ where $t = \sum_{i=1}^j t_i e_i$, we have

$$K_j^{(\alpha,\beta)}(z, z) = {}_3F_2\left(\frac{n}{r}, \frac{r}{2}a, (\alpha + \beta + (j-1)a + 2); \beta + \frac{j-1}{2}a + 1, \frac{j}{2}a; t^2\right).$$

We note that for a fixed α, $L_a^p\left(V_j, d\mu_j^{(\alpha,\beta)}\right)$ is independent of β.

Then the following result says that $K_j^{(\alpha,\beta)}(z,w)$ is the reproducing kernel of the Hilbert space $L_a^2\left(V_j, d\mu_j^{(\alpha,\beta)}\right)$.

THEOREM 2.20. *For* $f \in L_a^2\left(V_j, d\mu_j^{(\alpha,\beta)}\right)$, *then*

$$(24) \qquad f(z) = \int_{V_j} K_j^{(\alpha,\beta)}(z,w)f(w)d\mu_j^{(\alpha,\beta)}(w).$$

Again by using Theorem 1.9, we can obtain the asymptotics of $C_j(z,z)$ and $K_j^{(\alpha,\beta)}(z,z)$. In stead of stating the general results, we give the asymptotics in the special case when $j = 1$.

THEOREM 2.21. *As z varies in V_1,*

(i) $|C_1(z,z)| \approx (1 - |z|^2)^{-((r-1)a+b+1)}$;

(ii) $|K_1^{(0,\beta)}(z,z)| \approx (1 - |z|^2)^{-((r-1)a+b+2)}$.

We note that the dimension of V_1 equals $(r-1)a+b+1$. The Cauchy kernel $C(z,\zeta)$ and Bergman kernel $K(z,w)$ of the unit ball in $\mathbb{C}^{(r-1)a+b+1}$ are given by

$$C(z,\zeta) = (1 - \langle z,\zeta \rangle)^{-((r-1)a+b+1)}$$

and

$$K(z,w) = (1 - \langle z,w \rangle)^{-((r-1)a+b+2)}.$$

It is interesting to see from Theorem 2.21 that $C_1(z,z)$ and $K^{(0,\beta)}(z,z)$ have the same asymptotic behavior as that of the Cauchy kernel and the Bergman kernel of the unit ball in $\mathbb{C}^{(r-1)a+b+1}$.

REFERENCES

[A1]   J. Arazy, *Realization of the invariant inner products on the highest quotients of the composition series*, Arkiv for Matematik **30, No.1** (1992).

[A2]   ———, *Integral formulas for the invariant inner products in spaces of analytic functions on the ball*, Function Spaces (K. Jarosz, ed.), Lecture Notes in Pure and Applied Mathematics (Vol. 136), Marcel Dekker, 1992, pp. 9–23.

[CBZ] C. Berger, L. Coburn and K. Zhu, *BMO in the Bergman metric on bounded symmetric domains*, J. Funct. Anal. **93, No. 2** (1990), 310–350.

[DES] M. Davidson, T. Enright and R. Stanke, *Differential Operators and Highest Weight Representations*, Memoirs of AMS.

[F]   J. Faraut, *Algebres de Jordan et Cones Symetriques*, Ecole CIMPA, Université de Poitiers (1988).

[FK1] J. Faraut and A. Korányi, *Fonctions Hypergétriques associées aux cones symétriques*, C. R. Acad. Sci. Paris **301** (1988), 555–558.

[FK2] ———, *Function Spaces and Reproducing Kernels on Bounded Symmetric Domains*, J. Functional Analysis **88, No.1** (1990), 64–89.

[H]   S. Helgason, *Groups and Geometric Analysis*, Academic Press, New York, 1984.

[HU]   R. Howe and T. Umeda, *The Capelli Identity, the Double Commutant Theorem, and Multiplicity—Free Actions*, preprint.

[JK]   K. Johnson and A. Korányi, *The Hua operators and bounded symmetric domains of tube type*, Ann. Math. **111** (1980), 589–608.

[L]   M. Lassalle, *Les équations de Hua d'un domain borné symétrique du type tube*, Invent. Math. **77** (1984), 129–161.

[M]   C. C. Moore, *Compactification of Symmetric Spaces II. Cartan Domains*, Amer. J. Math. **86** (1964), 358–378.

[K1]   A. Korányi, *Complex Analysis and Symmetric Domains*, Ecole CIMPA—Université de Poitiers, 1988.

[K2]   _____, *Hua-type Integrals, Hypergeometric Functions and Symmetric Polynomials*, International Symposium in Memory of L. K. Hua, vol. 2, Springer-Verlag, pp. 169–180.

[KW]   A. Korányi and J. A. Wolf, *Realization of Hermitian Symmetric Spaces as Generalized Halfplanes*, Ann. of Math. **81** (1965), 165–288.

[M]   J. Mitchell, *Two-Sided L^p-Estimates ($p > 1$) for the Szego Kernel in Matrix Spaces with Application to a Mapping Theorem*, Complex Variables **18**, No.1–2 (1992), 73–78.

[MS]   J. Mitchell and G. Sampson, *Singular Integrals on Bounded Symmetric Domains in $\mathbb{C}^n$*, J. Math. Anal. and Appl. **90** (1982), 371–380.

[Mu]   R. J. Muirhead, *Aspects of Multivariate Statistical Theory*, J. Wiley, New York, 1982.

[N]   T. Nomura, *Algebraically independent Generators of Invariant Differential Operators on a Symmetric Cone*, J. Reine Angw. Math. **400** (1989), 122–133.

[S]   W. Schmid, *Die Randwerte holomorpher Funktionen auf hermitischen Raumen*, Invent. Math. **9** (1969), 61–80.

[U1]   H. Upmeier, *Jordan algebras and harmonic analysis on symmetric spaces*, Amer. J. Math. **108** (1986), 1–25.

[U2]   _____, *Fredholm Indices for Toeplitz Operators on Bounded Symmetric Domains*, Amer. J. Math. **110** (1988), 811–832.

[V]   M. Vergne and H. Rossi, *Analytic continuation of the holomorphic discrete series of a semi-simple Lie group*, Acta. Math. **136** (1976), 1–59.

[W]   N. Wallach, *The analytic continuation of the discrete series I, II*, Trans. Amer. Math. Soc. **251** (1979), 1–17, 19–37.

[Y1]   Z. Yan, *A Class of Generalized Hypergeometric Functions in Several Variables*, Can. J. Math. (to appear).

[Y2]   _____, *Generalized Hypergeometric Functions and Laguerre Polynomials in Two Variables*, Contemporary Math. (to appear).

[Y3]   _____, *Duality and differential operators on the Bergman spaces of bounded symmetric domains*, J. Functional Analysis **105**, No.1, April (1992), 171–186.

[Y4]   _____, *Invariant Differential Operators and Holomorphic Function Spaces on Bounded Symmetric Domains*, Priprint (1992).

[Y5]   _____, *Function Spaces on Varieties in a Bounded Symmetric Domain*, in preparation.

[Zh]   Dechao, Zheng, *Schatten Class Hankel operators on the Bergman Space*, Integral Equations Operator Theory **13**, No.3 (1990), 442–459.

[Z1]   K. Zhu, *Duality and Hankel operators on the Bergman spaces of bounded symmetric domains*, J. Functional Analysis **81** (1988), 260–278.

[Z2]   _____, *Holomorphic Besov Spaces on Bounded Symmetric Domains*, Preprint (1991).

[Z3]   _____, *Hankel operators on the Bergman space of bounded symmetric domains*, Trans. Amer. Math. Soc. (to appear).

DEPARTMENT OF MATHEMATICS, UNIVERSITY OF CALIFORNIA AT BERKELEY, BERKELEY, CA 94720

Contemporary Mathematics
Volume **142**, 1993

Beltrami Equation in High Dimensions

JINHAO ZHANG

The generalization of Teichmüller theory—infinitesimal deformation of compact complex manifolds developed by K. Kodaira and D. C. Spencer [5] depended on an overdetermined system of partial differential equations, which we call the Beltrami equation in high dimensions. It can be considered as a generalization of the Cauchy-Riemann equation and the Beltrami equation in one dimension. In §1, we show how the deformation of complex structures leads to this system.

The local theory of this system was established in the 1950's by the famous theorem of Newlander and Nirenberg. In studying the deformation of noncompact complex manifolds, one has to take further steps to investigate solutions of this system. We present the Neumann problem of Beltrami equation in high dimensions in §2, and show that the weak and strong pseudoconvexity can be linked by a homeomorphic solution of the Beltrami equation in high dimensions.

On the other hand, generalized homeomorphic solutions of the Beltrami equation in one dimension are a quasiconformal mappings. Conversely, any quasiconformal mapping satisfies a Beltrami equation. Quasiconformal mapping is a powerful tool in the study of complex structures on Riemann surfaces and, in particular, Teichmüller theory. As far as I know, contrary to the one dimensional case any generalization in high dimensions of quasiconformal mapping has no close link with complex structures. Nevertheless, if we proceed from the Beltrami equation in high dimensions as in §3, the generalization of quasiconformal mappings will fit in very naturally with complex structures.

This paper is based on the survey given by the author at the Beijing Conference in Several Complex Variables in May, 1990.

1. Generalization of Beltrami equation

Let X be a differentiable manifold of dimension $2n$. A complex structure $\{U_j, z_j\}$ on X consists of an open covering $\{U_j\}$ of X, and complex coordinate

1991 *Mathematics Subject Classification.* 32G05, 32-02.
Research supported by the National Science Foundation of China.
This paper is in final form and no version of it will be submitted for publication elsewhere.

charts z_j on U_j such that $z_j \circ z_k^{-1}$ is holomorphic whenever $U_j \cap U_k \neq \emptyset$. X provided with a complex structure $\{U_j, z_j\}$ is called a complex manifold.

Let $\{U_j, \zeta_j(z,t)\}$ be a family of complex structures depending on a parameter t, where z is a local complex coordinate on X. The mapping $\zeta_j(z,t) : U_j \to \mathbb{C}^n$, injective for any t, is differentiable on z, t, and is holomorphic in z when $t = 0$. For any $t = t_0$, we call $\{U_j, \zeta_j(z, t_0)\}$ a deformation of $\{U_j, z_j\}$. In the classical theory of infinitesimal deformations one needs to investigate the behavior of ζ_j when t is sufficiently small. Denote by $f_{jk} = \zeta_j \circ \zeta_k^{-1}(z,t)$ the coordinate transformation, then

$$\zeta_j(z,t) = f_{jk}(\zeta_k(z,t),t), \qquad z \in U_j \cap U_k,$$

or

$$(1) \qquad \zeta_j^\alpha(z,t) = f_{jk}^\alpha(\zeta_k^1(z,k), \ldots, \zeta_k^n(z,t), t), \qquad z \in U_j \cap U_k.$$

Taking $\bar{\partial}$, ∂ with respect to the complex structure $\{U_j, z_j\}$, and noting that f_{jk} is holomorphic in ζ_k, we obtain

$$(2) \qquad \begin{cases} \bar{\partial}\zeta_j^\alpha(z,t) = \sum \dfrac{\partial f_{jk}^\alpha}{\partial \zeta_k^\beta} \bar{\partial}\zeta_k^\beta(z,t), \\[2ex] \partial\zeta_j^\alpha(z,t) = \sum \dfrac{\partial f_{jk}^\alpha}{\partial \zeta_k^\beta} \partial\zeta_k^\beta(z,t). \end{cases}$$

Denote by $\left(\dfrac{\partial \zeta_j}{\partial z} \right)$, $\left(\dfrac{\partial \zeta_j}{\partial \bar{z}} \right)$, $\left(\dfrac{\partial f_{jk}}{\partial \zeta_k} \right)$ the matrices

$$\begin{pmatrix} \dfrac{\partial \zeta_j^1}{\partial z^1} & \cdots & \dfrac{\partial \zeta_j^n}{\partial z^1} \\ \vdots & \ddots & \vdots \\ \dfrac{\partial \zeta_j^1}{\partial z^n} & \cdots & \dfrac{\partial \zeta_j^n}{\partial z^n} \end{pmatrix}, \quad \begin{pmatrix} \dfrac{\partial \zeta_j^1}{\partial \bar{z}^1} & \cdots & \dfrac{\partial \zeta_j^n}{\partial \bar{z}^1} \\ \vdots & \ddots & \vdots \\ \dfrac{\partial \zeta_j^1}{\partial \bar{z}^n} & \cdots & \dfrac{\partial \zeta_j^n}{\partial \bar{z}^n} \end{pmatrix}, \quad \begin{pmatrix} \dfrac{\partial f_{jk}^1}{\partial \zeta_k^1} & \cdots & \dfrac{\partial f_{jk}^n}{\partial \zeta_k^1} \\ \vdots & \ddots & \vdots \\ \dfrac{\partial f_{jk}^1}{\partial \zeta_k^n} & \cdots & \dfrac{\partial f_{jk}^n}{\partial \zeta_k^n} \end{pmatrix}$$

respectively, (2) can be written as

$$(3) \qquad \left(\dfrac{\partial \zeta_j}{\partial \bar{z}} \right) = \left(\dfrac{\partial \zeta_k}{\partial \bar{z}} \right) \left(\dfrac{\partial f_{jk}}{\partial \zeta_k} \right), \qquad \left(\dfrac{\partial \zeta_j}{\partial z} \right) = \left(\dfrac{\partial \zeta_k}{\partial z} \right) \left(\dfrac{\partial f_{jk}}{\partial \zeta_k} \right).$$

At $t = 0$, $(z^1, \ldots, z^n)$ are local complex analytic coordinates on X and $(\zeta_j^1(z,0), \ldots, \zeta_j^n(z,0))$ are also local complex analytic, so we have

$$\det \left(\dfrac{\partial \zeta_j(z,0)}{\partial z} \right) \neq 0.$$

Hence for $|t|$ small enough,

$$\det \left(\dfrac{\partial \zeta_j(z,t)}{\partial z} \right) \neq 0.$$

Therefore, by (3)

$$(4) \qquad \left(\frac{\partial \zeta_j}{\partial \bar{z}}\right)\left(\frac{\partial \zeta_j}{\partial z}\right)^{-1} = \left(\frac{\partial \zeta_k}{\partial \bar{z}}\right)\left(\frac{\partial \zeta_k}{\partial z}\right)^{-1}.$$

Denote by M the matrix

$$M = \begin{pmatrix} \mu_1^1 & \cdots & \mu_1^n \\ \vdots & \ddots & \vdots \\ \mu_n^1 & \cdots & \mu_n^n \end{pmatrix} = \left(\frac{\partial \zeta_j}{\partial \bar{z}}\right)\left(\frac{\partial \zeta_j}{\partial z}\right)^{-1}.$$

Then we have

$$\frac{\partial \zeta_j}{\partial \bar{z}} = M\left(\frac{\partial \zeta_j}{\partial z}\right),$$

or

$$(5) \qquad \frac{\partial \zeta_j^\alpha}{\partial \bar{z}^\beta} - \sum_\gamma \mu_\beta^\gamma \frac{\partial \zeta_j^\alpha}{\partial z^\gamma} = 0, \qquad \alpha, \beta = 1, \ldots, n.$$

PROPOSITION 1. *Suppose that* $\det\left(\dfrac{\partial \zeta^\alpha}{\partial z^\gamma}\right) \neq 0.$ *Then a function* f *defined on* X *is holomorphic in* ζ *if and only if* f *satisfies the system of equations*

$$(6) \qquad \frac{\partial f}{\partial \bar{z}^\beta} - \sum_\gamma \mu_\beta^\gamma \frac{\partial f}{\partial z^\gamma} = 0, \qquad \beta = 1, \ldots, n.$$

For the proof please see [9]. The system of partial differential equations (6) is overdetermined, which we call a Beltrami equation in high dimensions. When $\mu_\beta^\gamma \equiv 0$, $\beta, \gamma = 1, \ldots, n$, (6) reduces to the Cauchy-Riemann equation, and when $n = 1$, (6) reduces to the Beltrami equation in one dimension.

By virtue of (4), the matrix M corresponds to a global vector-valued $(0, 1)$-form μ on X:

$$\mu = \sum_{\beta,\gamma} \mu_\beta^\gamma(z, t)d\bar{z}^\beta \otimes \frac{\partial}{\partial z^\gamma},$$

which can stand for the complex structure $\{U_j, \zeta_j(z, t)\}$.

In fact, any global vector-valued $(0, 1)$-form represents an almost complex structure on X. We say that an almost complex structure μ is integrable if μ satisfies the following condition:

$$(7) \qquad \frac{\partial \mu_\beta^\alpha}{\partial \bar{z}^\gamma} - \frac{\partial \mu_\gamma^\alpha}{\partial \bar{z}^\beta} = \sum\left(\mu_\gamma^\sigma \frac{\partial \mu_\beta^\alpha}{\partial z^\sigma} - \mu_\beta^\sigma \frac{\partial \mu_\gamma^\alpha}{\partial z^\sigma}\right), \qquad \alpha, \beta, \gamma = 1, \ldots, n.$$

In 1957, by using the integral representation of solutions of the $\bar{\partial}$-equation on polydisks, Newlander and Nirenberg proved the following theorem [7]:

THEOREM 2. *If μ satisfies (7) and the system of equations (6) is linearly independent, then (6) has a local holomorphic solution.*

Obviously, (6) is linearly independent when $|t|$ is sufficiently small. If we denote by $\zeta_j = (\zeta_j^1, \ldots, \zeta_j^n)$ the holomorphic solution of (6) on U_j, then $\det\left(\dfrac{\partial \zeta^\alpha}{\partial z^\gamma}\right) \neq 0$ when t is small enough. Therefore from proposition 1 we have

COROLLARY 3. *For small enough t, the integrable almost complex structure determines a complex structure.*

Another proof of theorem 2 were provided by Kohn [4] and Hörmander [3]. Malgrange [6] has obtained a very simple proof of this result, which is applicable in much more general situations.

2. $_\mu\bar\partial$-Neumann Problem

In this section the almost complex structure $\mu = \mu(z)$ is independent of any parameters. Suppose that μ satisfies the following three conditions:

(I) integrability condition;
(II) $\det(I - M\bar M) \neq 0$, where M is the matrix $\left(\mu_j^k\right)$;
(III) $\mu_j^k \in C^\infty(X)$, $j, k = 1, \ldots, n$.

Note that condition (II) is independent of local coordinates.

Because of condition (II), the Beltrami equation in high dimensions (6) is linear independent. It follows from theorem 2 that a holomorphic solution $\zeta = (\zeta^1, \ldots, \zeta^n)$ existed. Let J be the Jacobian matrix of ζ. Noting that ζ^α are solutions of (6), one can see

$$(8) \qquad \det J = \det(I - M\bar M)\left|\det\left(\frac{\partial \zeta}{\partial z}\right)\right|^2.$$

By (8) $\det\left(\dfrac{\partial \zeta}{\partial z}\right) \neq 0$, when $\det J \neq 0$ and $\det(I - M\bar M) \neq 0$. From proposition 1 we obtain

PROPOSITION 4. *Any almost complex structure satisfying conditions (I), (II), (III) determines a complex structure on X.*

Given an almost complex structure μ on X, we can define an operator $_\mu\bar\partial : C_{p,q}^\infty(X) \to C_{p,q+1}^\infty(X)$ as follows. Denote

$$\mu^\beta = \sum_\gamma \mu_\gamma^\beta d\bar z^\gamma, \qquad \partial_\beta = \frac{\partial}{\partial z^\beta}.$$

For any (p,q)-form $\phi \in C_{p,q}^\infty(X)$, locally

$$\phi = \sum_{I,J} \phi_{I,J} dz^I \wedge d\bar z^J.$$

Define $_\mu\bar\partial\phi$ by

$$_\mu\bar\partial\phi = \bar\partial\phi - \sum\mu^\beta \wedge \partial_\beta\phi$$
$$= \sum_{I,J}\sum_{j=1}^{n}\left(\bar\partial_j\phi_{I,J} - \sum_{k=1}^{n}\mu_j^k\partial_k\phi_{I,J}\right)d\bar z^j \wedge dz^I \wedge d\bar z^J.$$

With the operator $_\mu\bar\partial$, we can write the integrability condition of μ as

$$_\mu\bar\partial\mu^\alpha = 0, \qquad \forall\alpha,$$

and the Beltrami equation in high dimensions as

$$(9) \qquad\qquad _\mu\bar\partial f = 0.$$

A function f satisfying (9) is called a μ-holomorphic function. A μ-holomorphic function is holomorphic with regard to the complex structure μ. Therefore, μ-holomorphic functions assume many properties of holomorphic functions. For example,

   (a) The maximum principle for μ-holomorphic functions also holds.
   (b) The Hartogs phenomena of extension for μ-holomorphic functions also appear.
   (c) If two μ-holomorphic functions f, g are equal on an open set of X, then they are identical on X.
   (d) The zero set of a μ-holomorphic function is not isolated.

Now we consider the $_\mu\bar\partial$-Neumann problem. For simplicity, we assume that X is a bounded domain $D \subset \mathbb{C}^n$ with smooth boundary. Let $_\mu\bar\partial^*$ be the L^2-adjoint of $_\mu\bar\partial$. Denote by $\mathrm{Dom}(_\mu\bar\partial)$, $\mathrm{Dom}(_\mu\bar\partial^*)$ the defining domains of $_\mu\bar\partial$, $_\mu\bar\partial^*$ respectively. Set

$$H_{p,q} = \left\{\phi \in \mathrm{Dom}(_\mu\bar\partial)\cap\mathrm{Dom}(_\mu\bar\partial^*) \mid {_\mu\bar\partial}\phi = 0, {_\mu\bar\partial^*}\phi = 0\right\}.$$

The $_\mu\bar\partial$-Neumann problem for (p,q)-forms can be stated as follows: Given $\theta \in L^2_{p,q}(D)$ with $\theta \perp H_{p,q}$, does there exist a unique form $\phi \in \mathrm{Dom}(_\mu\bar\partial)\cap\mathrm{Dom}(_\mu\bar\partial^*)$ with $_\mu\bar\partial\phi \in \mathrm{Dom}(_\mu\bar\partial^*)$ and $_\mu\bar\partial^*\phi \in \mathrm{Dom}(_\mu\bar\partial)$ such that

$$(10) \qquad\qquad _\mu\bar\partial{_\mu\bar\partial^*}\phi + {_\mu\bar\partial^*}{_\mu\bar\partial}\phi = \theta?$$

In addition, what can one say about the regularity of ϕ?

Let

$$_\mu\bar\partial_j = \bar\partial_j - \sum_k\mu_j^k\partial_k, \qquad _\mu\partial_j = \partial_j - \sum_k\bar\mu_j^k\bar\partial_k.$$

From condition (II) we can find $a_j^k, b_j^k \in C^\infty(\bar D)$ such that

$$\partial_j = \sum_k\left(a_j^k\,{_\mu\partial_k} + b_j^k\,{_\mu\bar\partial_k}\right),$$
$$\bar\partial_j = \sum_k\left(\bar a_j^k\,{_\mu\bar\partial_k} + \bar b_j^k\,{_\mu\partial_k}\right).$$

Proceeding as in dealing with the $\bar\partial$-Neumann problem, we proved in [9] the following:

THEOREM 5. *Let $D \in \mathbb{C}^n$ be a bounded domain with smooth boundary, and μ a C^∞ tensor field defined on a neighborhood of $\bar{D}$ satisfying the conditions (I), (II) and (III). And for any $\xi = (\xi^1, \ldots, \xi^n) \in \mathbb{C}^n$, $\xi \neq 0$,*

$$
\begin{aligned}
(11) \quad \sum_{i,j} \Bigg[&\partial_i \bar{\partial}_j r - \sum_k \mu_j^k \partial_i \partial_k r - \sum_\ell \bar{\mu}_i^\ell \bar{\partial}_\ell \bar{\partial}_j r + \sum_{\ell,k} \bar{\mu}_i^\ell \mu_j^k \bar{\partial}_\ell \partial_k r \\
&- \sum_{\ell,t} \left({}_\mu \bar{\partial}_\ell r \right) \left({}_\mu \bar{\partial}_j \bar{\mu}_i^t \right) \bar{a}_t^\ell - \sum_{\ell,t} \left({}_\mu \partial_\ell r \right) \left({}_\mu \partial_i \mu_j^t \right) a_t^\ell \Bigg] \xi^i \bar{\xi}^j > C' |\xi|^2 \\
&\forall \sum_j {}_\mu \partial_j r \xi^j = 0, \qquad z \in bD,
\end{aligned}
$$

then there exists a unique ϕ satisfying (10) for any $\theta \perp H_{p,q}$, and $\phi \in C_{p,q}^\infty(\bar{D})$ whenever $\theta \in C_{p,q}^\infty(\bar{D})$.

It is interesting to ask the following question:

QUESTION 1. *For what condition on D and μ, can the ${}_\mu\bar{\partial}$-Neumann problem be solved?*

Noting the inequality (11) only concerns the value of μ_j^k and its first derivatives on bD, we have

COROLLARY 6. *Suppose that D is a strongly pseudoconvex domain. Let μ be defined on a neighborhood of $\bar{D}$ satisfying conditions (I), (II), (III) and be sufficiently small on bD (boundary of D). Then the ${}_\mu\bar{\partial}$-Neumann problem of D can be solved.*

For a general domain, the case is very complicated. It is well-known that the $\bar{\partial}$-Neumann problem can be solved on a strongly pseudoconvex domain or pseudoconvex domain of finite type (see [2], [4]). So a natural question is how to characterize a domain on which the ${}_\mu\bar{\partial}$-Neumann problem can be solved.

THEOREM 7. *Let D, μ be defined as in theorem 5. Then for a given point $p \in bD$, any 1-1 μ-holomorphic mapping $F = (f_1, \ldots, f_n)$ defined on a neighborhood U of p takes p to a strongly pseudoconvex point of $F(U \cap bD)$.*

This theorem is applicable to the weak version of question 1 as follows.

QUESTION 1'. *For a given domain D and a point $p \in bD$, does there exist a μ defined on a neighborhood of p satisfying conditions (I), (II), (III) such that inequality (11) holds in $U \cap bD$?*

In [8] we gave an affirmative answer of question 1'. So from theorem 7 we have

COROLLARY 8. *Every sufficiently small piece of smooth boundary of a domain is strongly pseudoconvex under a suitable local complex structure.*

If the domain is pseudoconvex, we can say more:

COROLLARY 9. *Every sufficiently small piece of smooth boundary of a pseudoconvex domain is strongly pseudoconvex under a suitable local complex structure, which is sufficiently close to the original complex structure.*

QUESTION 2. *Given a domain D, suppose that the $_\mu\bar\partial$-Neumann problem can be solved for some μ. Does there exist a 1-1 μ-holomorphic mapping on D globally?*

In the case of $n = 1$, the answer of question 2 is affirmative when $|\mu| < 1$. It has many important applications in the theory of quasiconformal mappings. For the answer of question 2 in general, we need to understand the Beltrami equation in high dimensions and the system of integrability (7) more deeply.

On the other hand, if there is a 1-1 μ-holomorphic mapping of D, on which the $_\mu\bar\partial$-Neumann problem can be solved, then D is μ-holomorphically homeomorphic to a strongly pseudoconvex domain, or in other words D is strongly pseudoconvex from the view point of the complex structure μ.

3. Quasiconformal mappings in high dimensions

In the one dimensional case, homeomorphic solutions of the Beltrami equation are quasiconformal mappings. However, homeomorphic solutions of the Beltrami equation in high dimensions is not necessarily quasiconformal, even for biholomorphic mappings, the homeomorphic solutions of Cauchy-Riemann equation.

QUESTION 3. *What is the relation between quasiconformal mappings and the Beltrami equation in high dimensions?*

Let $(x_1, \ldots, x_{2n})$ be the real coordinates of $\mathbb{R}^{2n}$, and $F = (y_1, \ldots, y_{2n})$ a differentiable mapping on $\mathbb{R}^{2n}$. It is well-known that F is K-quasiconformal if and only if the inequality

$$\sum_{j,k} \left| \frac{\partial y_j}{\partial x_k} \right|^2 \leq K \left| \det \left(\frac{\partial y_j}{\partial x_k} \right) \right|^{\frac{2}{n}}$$

holds.

For $\mathbb{R}^{2n}$ we introduce a natural complex structure by letting $z^j = x_j + ix_{n+j}$, $j = 1, \ldots, n$. Denote by $f_j = y_j + iy_{n+j}$, $j = 1, \ldots, n$. Let A, B be the matrices

$$A = \left(\frac{\partial f_j}{\partial z^k} \right)_{j,k=1,\ldots,n}, \qquad B = \left(\frac{\partial f_j}{\partial \bar z^k} \right)_{j,k=1,\ldots,n},$$

respectively. Furthermore, suppose that $\det A \neq 0$. Let

$$M = BA^{-1} = \left(\mu_j^k \right).$$

Then $\mu = \sum \mu_j^k d\bar z^j \otimes \frac{\partial}{\partial z^k}$ defines a new complex structure on $\mathbb{R}^{2n}$. We have

PROPOSITION 10. $F = (f_1, \ldots, f_n)$ *is a holomorphic solution of the Beltrami equation in high dimensions with regard to this μ.*

On the other hand, instead of condition (II) we introduce condition (II'):

(II') $\sum_{j,k} |\mu_j^k|^2 < K < 1$.

Note that the condition (II') is not biholomorphically invariant and depends on the choice of coordinates in the same complex structure, while it is conformally invariant.

THEOREM 11. *Let* $\mu = (\mu_j^k)$ *satisfy conditions (I), (II'), (III) and let* $F = (f_1, \ldots, f_n)$ *be a global homeomorphic solution of Beltrami equation in high dimensions such that*

$$\sum_{\ell,k} \left| \frac{\partial f_\ell}{\partial z^k} \right|^2 \leqslant C \left| \det \left(\frac{\partial f_\ell}{\partial z^k} \right) \right|^{\frac{2}{n}}.$$

Then F *is a* K*-quasiconformal mapping.*

For the proof and applications of this theorem, please see [10].

REFERENCES

1. L. V. Ahlfors, *Lectures on Quasiconformal Mappings*, Van Nostrand, New York, 1966.
2. D. Catlin, *Subelliptic estimates for the $\bar{\partial}$-Neumann problem on pseudoconvex domains*, Ann. of Math. **126** (1987), 131–191.
3. L. Hörmander, *The Frobenius-Nirenberg theorem*, Arkiv. Mat. **5** (1965), 425–432.
4. J. J. Kohn, *Harmonic integrals on strongly pseudoconvex manifolds. I*, Ann. of Math. **78** (1963), 112–148; II, ibid. **79** (1964), 450–472.
5. K. Kodaira, *Complex Manifolds and Deformation of Complex Structure*, Springer-Verlag, 1986.
6. B. Malgrange, *Pseudo-groupes de Lie elliptiques*, Séminaire Leray, Collège de France, 1960–70.
7. A. Newlander and L. Nirenberg, *Complex analytic coordinates in almost-complex manifolds*, Ann. of Math. **65** (1957), 391–404.
8. Xuan Ye and Jinhao Zhang, *Existence of μ-holomorphic separating functions of bounded smooth domains* (to appear).
9. Jinhao Zhang, *The Neumann problem for an overdetermined system related with almost complex structure*, Science Sinica (to appear).
10. ———, *Quasiconformal mapping and Beltrami equation in high dimension* (to appear).

DEPARTMENT OF MATHEMATICS, FUDAN UNIVERSITY, SHANGHAI 200433, PEOPLE'S REPUBLIC OF CHINA

Contemporary Mathematics
Volume **142**, 1993

Singular Integrals and Integral Representations in Several Complex Variables

TONGDE ZHONG

Contents

I. Plemelj formulas with Bochner-Martinelli kernel and their applications

1. Plemelj formula with Bochner-Martinelli kernel.

1991 *Mathematics Subject Classification.* 32A25, 32A40, 32-02.
Research supported by the National Science Foundation of China.
This paper is in final form and no version of it will be submitted for publication elsewhere.

It is well-known that in one complex variable Plemelj formula plays an important role in singular integral equations. One may ask in several complex variables case whether there exists a Plemelj formula that plays a similar role as in one complex variable. In 1957, Lu Qikeng and Zhong Tongde [**LZ**] obtained a Plemelj formula with Bochner-Martinelli kernel in $\mathbb{C}^n$.

Let D be a bounded domain in $\mathbb{C}^n$, its boundary ∂D be a smooth orientable manifold of dimension $2n - 1$ of class C^2. Let $S_\delta(\zeta)$ be a ball centered at $\zeta \in \partial D$ with radius δ. Denote the inner part of the boundary ∂D in $S_\delta(\zeta)$ by $\sigma_\delta(\zeta, \xi) = \partial D_\zeta \cap S_\delta(\zeta)$ and the remainder of the boundary by $\Sigma_\delta(\zeta, \xi) = \partial D \backslash \sigma_\delta(\zeta, \xi)$.

A function $f(\xi)$ defined on the boundary ∂D is said to satisfy Hölder condition on the boundary, denoted $f(\xi) \in H(\alpha, \partial D)$, if for two arbitrary points $\xi, \eta \in \partial D$, the inequality

$$(1.1) \qquad |f(\xi) - f(\eta)| \leqslant A|\xi - \eta|^\alpha$$

holds for a positive constant A, called the Hölder constant, and a constant α, $0 < \alpha \leqslant 1$, called the Hölder exponent.

The well-known Bochner-Martinelli kernel in $\mathbb{C}^n$ is expressed by

$$(1.2) \qquad K(\zeta, z) \equiv \Omega(\bar{\zeta} - \bar{z}, \zeta - z) := \frac{(n-1)!}{(2\pi i)^n} \frac{\omega'(\bar{\zeta} - \bar{z}) \wedge \omega(\zeta - z)}{\langle \bar{\zeta} - \bar{z}, \zeta - z \rangle^n},$$

where

$$\omega'(v) := \sum_{j=1}^{n} (-1)^{j+1} v_j \, dv_1 \wedge \cdots \wedge \widehat{dv_j} \wedge \cdots \wedge dv_n,$$

$$\omega(v) := dv_1 \wedge \cdots \wedge dv_n,$$

$$\langle v, u \rangle := \sum_{j=1}^{n} v_j u_j.$$

Let $f(\xi)$ be a function defined on ∂D, the integral

$$(1.3) \qquad F(\eta) = \int_{\xi \in \partial D} f(\xi) K(\xi, \eta), \qquad \eta \in \partial D$$

is a singular integral on ∂D. The principal value of this singular integral is defined as

$$(1.4) \qquad \mathrm{V.\,P.}\, F(\eta) = \lim_{\delta \to 0} \int_{\xi \in \partial D \backslash \sigma_\delta} f(\xi) K(\xi, \eta), \qquad \eta \in \partial D.$$

Lu Qikeng and Zhong Tongde [**LZ**] proved that if $f(\xi) \in H(\alpha, \partial D)$, then the principal value (1.4) exists.

Furthermore we proved

THEOREM 1.1 (PLEMELJ FORMULA). *If ∂D is a smooth orientable manifold of class C^2, $f(\xi)$ is a continuous complex valued function defined on ∂D and $f(\xi) \in H(\alpha, \partial D)$, then for the Cauchy type integral*

$$(1.5) \qquad F(z) = \int_{\partial D} f(\xi) K(\xi, z), \qquad z \in D^+,$$

we have

$$(1.6) \qquad F^+(\zeta) = \text{V. P.} \int_{\partial D} f(\xi) K(\xi, \zeta) + \frac{1}{2} f(\zeta), \qquad \zeta \in \partial D,$$

$$(1.7) \qquad F^-(\zeta) = \text{V. P.} \int_{\partial D} f(\xi) K(\xi, \zeta) - \frac{1}{2} f(\zeta), \qquad \zeta \in \partial D,$$

where $F^+(\zeta)$ and $F^-(\zeta)$ denote the limit values of $F(z)$ when z approaches $\zeta \in \partial D$ from D^+ and D^- respectively.

From (1.6) and (1.7) we have immediately the Plemelj jump formula

$$(1.8) \qquad F^+(\zeta) - F^-(\zeta) = f(\zeta), \qquad \zeta \in \partial D$$

and $F^{\pm}(z) \in H(\alpha, \partial D)$ (V. A. Kakichev [**K**]).

Plemelj formula can also be extended to Stein manifold.

Let M be a Stein manifold, $T(M)$ the complex tangent bundle of M, and $\tilde{T}(M \times M)$ the pull-back of $T(M)$ relative to the projection $M \times M \to M$, $(z, \zeta) \to z$. The section $s(z, \zeta)$ generates the ideal of $\Delta(M)$ in ${}_{M \times M}\mathcal{O}$. Then for a relative compact domain D of M we have (Lin Yaxian [**LinY**], Zhong Tongde [**Z7**])

THEOREM 1.2 (PLEMELJ FORMULA). *Suppose that the boundary ∂D of D is a smooth orientable manifold of class C^2. We can define a Bochner-Martinelli type integral*

$$(1.9) \qquad \int_{\partial D} f(\zeta) \Omega^0(\phi^{\nu}, \bar{s}, s; z) = \begin{cases} F^+(z), & z \in D := D^+, \\ F^-(z), & z \in M \backslash \bar{D} := D^-, \end{cases}$$

where

$$(1.10) \qquad \Omega^0(\phi^{\nu}, \bar{s}, s; z) = \frac{(n-1)!}{(2\pi i)^n} \phi^{\nu n}(z, \zeta) \omega'\left(\frac{\bar{s}(z, \zeta)}{|s(z, \zeta)|^2}\right) \wedge \omega(s(z, \zeta)),$$

$\phi(z, \zeta)$ is a holomorphic function on $M \times M$ such that for all $z \in M$, $\phi(z, z) = 1$. For sufficiently large integer ν, $\Omega^0(\ )$ is a continuous form on $M \backslash \{z\}$. If $f(\zeta)$ is a continuous complex-valued function defined on ∂D, and $f(\zeta) \in H(\alpha, \partial D)$, $0 < \alpha < 1$, that is, for two arbitrary point $\xi, \zeta \in \partial D$, the inequality

$$(1.11) \qquad |f(\xi) - f(\zeta)| \leqslant A[d(\xi, \zeta)]^{\alpha}, \qquad 0 < \alpha < 1$$

holds, where A is a positive constant. Then the following Plemelj formula

(1.12)
$$F^+(\eta) = \mathrm{V.P.} \int_{\partial D} f(\zeta)\Omega^0(\phi^\nu, \bar{s}, s; \eta) + \frac{1}{2}f(\eta), \qquad \eta \in \partial D,$$

(1.13)
$$F^-(\eta) = \mathrm{V.P.} \int_{\partial D} f(\zeta)\Omega^0(\phi^\nu, \bar{s}, s; \eta) - \frac{1}{2}f(\eta), \qquad \eta \in \partial D,$$

are valid, where $F^+(\eta)$ and $F^-(\eta)$ denote the limits of $F^+(z)$ and $F^-(z)$ when z approaches $\eta \in \partial D$ from D^+ and D^- respectively, and

(1.14)
$$\mathrm{V.P.} \int_{\zeta \in \partial D} f(\zeta)\Omega^0(\phi^\nu(\eta, \zeta), \bar{s}(\eta, \zeta), s(\eta, \zeta); \eta)$$
$$= \int_{\zeta \in \partial D \setminus V_{\eta\delta}} f(\zeta)\Omega^0(\phi^\nu(\eta, \zeta), \bar{s}(\eta, \zeta), s(\eta, \zeta); \eta)$$
$$+ \int_{\zeta^* \in H} \hat{f}(\zeta^*)\hat{\phi}^{\nu_n}(0, \zeta^*)A(0, \zeta^*)$$
$$+ \int_{\zeta^* \in H} \hat{f}(\zeta^*)\hat{\phi}^{\nu_n}(0, \zeta^*)[B(0, \zeta^*) - \Omega(\bar{\zeta}^* - 0, \zeta^* - 0)]$$
$$+ \int_{\zeta^* \in H} \hat{f}(\zeta^*)[\hat{\phi}^{\nu_n}(0, \zeta^*) - 1]\Omega(\bar{\zeta}^* - 0, \zeta^* - 0)$$
$$+ \lim_{\varepsilon \to 0} \int_{\substack{\zeta^* \in H \\ |\zeta^*| > \varepsilon}} \hat{f}(\zeta^*)\Omega(\bar{\zeta}^* - 0, \zeta^* - 0),$$

where for fixed $\eta \in \partial D$, take $\delta > 0$, such that $W_\eta = s(\eta, \cdot) : V_{\eta\delta} \to B_\delta = \{\zeta^* \in \mathbb{C}^n \mid |\zeta^*| < \delta\}$ is a biholomorphic map, $V_{\eta\delta}$ is a small neighborhood of η in the coordinate neighborhood, $z = W_\eta^{-1}(z^*)$, $\zeta = W_\eta^{-1}(\zeta^*)$,

(1.15)
$$\hat{\bar{s}}(z^*, \zeta^*) = \bar{s}(W_\eta^{-1}(z^*), W_\eta^{-1}(\zeta^*)),$$
$$\hat{s}(z^*, \zeta^*) = s(W_\eta^{-1}(z^*), W_\eta^{-1}(\zeta^*)),$$
$$\hat{\phi}(z^*, \zeta^*) = \phi(W_\eta^{-1}(z^*), W_\eta^{-1}(\zeta^*)),$$
$$\hat{f}(\zeta^*) = f(W_\eta^{-1}(\zeta^*)).$$

Let $\{u_j\}$, $\{\bar{u}_j\}$ be the coordinates of $\hat{s}$, $\hat{\bar{s}}$, then $\{u_j\}$ are holomorphic functions on $B_\delta \times B_\delta$ and can be expressed by

(1.16)
$$u_j(z^*, \zeta^*) = \sum_{k=1}^n \gamma_{jk}(z^*, \zeta^*)(\zeta_k^* - z_k^*).$$

Moreover

(1.17)
$$B(z^*, \zeta^*) = \frac{(n-1)!}{(2\pi i)^n} \frac{\det(\bar{\Gamma}_t \Gamma)\omega'(\bar{\zeta}^* - \bar{z}^*) \wedge \omega(\zeta^* - z^*)}{\left[\sum_{jk\ell} \gamma_{jk}\bar{\gamma}_{k\ell}(\bar{\zeta}_\ell^* - \bar{z}_\ell^*)(\zeta_j^* - z_j^*)\right]^n},$$

$\Gamma = (\gamma_{jk}(z^*, \zeta^*))$. We have $F^\pm(z) \in H(\alpha, \partial D)$, besides, if $\partial D \in C^k$ ($1 \leqslant k \leqslant \infty$) and $f \in C^\gamma(\partial D)$, $0 \leqslant \gamma \leqslant k$, then $F^+(z) \in C^\gamma(\bar{D}^+)$, $F^-(z) \in C^\gamma(\bar{D}^-)$.

From (1.12) and (1.13) we have the Plemelj jump formula

(1.18) $F^+(\eta) - F^-(\eta) = f(\eta), \qquad \eta \in \partial D,$

immediately.

2. Singular integral equations on smooth closed manifolds.

2.1 Composite formula and singular integral equations with constant coefficients.

Suppose that $f(\zeta) \in C^{(1)}(\partial D)$, and that it can be holomorphically extended into D, then

(2.1) $$f(z) = \int_{\zeta \in \partial D} f(\zeta) K(\zeta, z), \qquad z \in D$$

and

(2.2) $$F^+(\eta) = f(\eta), \qquad \eta \in \partial D.$$

If we introduce singular integral operator

(2.3) $$Sf = 2 \int_{\zeta \in \partial D} f(\zeta) K(\zeta, \eta), \qquad \eta \in \partial D,$$

then the following composite formula

(2.4) $$S^2 f = f$$

holds.

Let $\mathcal{L}$ denote the linear space consisting of all complex valued continuous functions which belong to C^1 on ∂D. Consider singular integral equation

(2.5) $$af + bSf + Tf = g$$

where a, b are complex constants satisfying $a^2 - b^2 \neq 0$. Then given $f \in \mathcal{L}$ which can be extended holomorphically into D, and

$$Tf = \int_{\xi \in \partial D} f(\xi) L(\xi, \eta)$$

where kernel $L(\xi, \eta)$ is a complex exterior differential form of degree $2n - 1$, which belong to C^1 with respect to ξ, η.

By composite formula (2.4) we have

THEOREM 2.1. *Suppose that in equation (2.4), a, b are complex constants, satisfying $a^2 - b^2 \neq 0$. Given $f \in \mathcal{L}$ that can be extended holomorphically into D and assume that the kernel $L(\xi, \eta)$ of integral operator T belongs to C^1, then*

(i) *the characteristic equation of (2.5)*

(2.6) $$af + bSf = g$$

has unique solution:

(2.7)
$$f = \frac{1}{a+b}g$$

in $\mathcal{L}$.

(ii) *Singular integral equation (2.5) is equivalent to Fredholm equation*

(2.8)
$$a_1 f + T^* f = F$$

in $\mathcal{L}$, where $a_1 = a^2 - b^2 \neq 0$, $F = (aI - bS)g$, $T^* = (aI - bS)T$.

2.2 Transformation formula and singular integral equation with variable coefficients.

THEOREM 2.2 (TRANSFORMATION FORMULA). *If $f(\eta, \xi) \in H(\alpha, \partial D)$, $\zeta \in \partial D$, then*

(2.9)
$$\int_{\eta \in \partial D} K(\eta, \zeta) \int_{\xi \in \partial D} f(\eta, \xi) K(\xi, \eta)$$
$$= \int_{\xi \in \partial D} \int_{\eta \in \partial D} f(\eta, \xi) K(\eta, \zeta) K(\xi, \eta) + \frac{1}{4} f(\zeta, \zeta),$$

where the two fold integrals on the left hand side take Cauchy principal value according to (1.4), the inner integral of the first term on the right hand side takes the principal value

(2.10)
$$\text{V.P.} \int_{\eta \in \partial D} f(\eta, \xi) K(\eta, \zeta) K(\xi, \eta)$$
$$= \lim_{\delta \to 0} \int_{\Sigma_\delta(\xi, \eta) \cap \Sigma_\delta(\zeta, \eta)} f(\eta, \xi) K(\eta, \zeta) K(\xi, \eta),$$

and the outer integral is an ordinary integral (Sun Jiguang [S], Zhong Tongde [Z3]).

Now given $f \in \mathcal{L}$ and suppose that f can be holomorphically extended into D. We discuss the singular integral equation

(2.11)
$$Rf \equiv af + bSf + Tf = g$$

and solve it in $\mathcal{L}$, where $a(\eta)$, $b(\eta)$, and $g(\eta)$ belong to $\mathcal{L}$. Let operator L be defined as

(2.12)
$$Lf = \int_{\xi \in \partial D} f(\xi) L(\xi, \eta),$$

where $L(\xi, \eta)$ is a complex exterior differential form of degree $2n - 1$ on ∂D,

(2.13) $L(\xi, \eta) = \displaystyle\sum_{k=1}^{n} \frac{\ell_k(\xi, \eta)}{|\xi - \eta|^{2n-1-\gamma}} \bar{d}\xi_1 \wedge d\xi_1 \wedge \cdots \wedge [\bar{d}\xi_k] \wedge d\xi_k \wedge \cdots \wedge \bar{D}\xi_n \wedge d\xi_n,$

where

(2.14) $\gamma > 0,$ $\ell_k(\xi, \eta) \in H(\alpha, \partial D),$ $0 < \alpha < 1,$ $1 \leqslant k \leqslant n.$

Using transformation formula (2.9), Sun Jiguang [S] has proved the following:

THEOREM 2.3. *If f, a, b, s, L and g in equation (2.10) are described as above, and $a^2 - b^2 \neq 0$ on ∂D, then (2.10) can be reduced to a Fredholm type equation. Especially, if a and b are constants, then (2.11) is equivalent to a Fredholm type equation, and its characteristic equation (2.6) has unique solution (2.7) in $\mathcal{L}$.*

The above results about Plemelj formula, composite formula, transformation formula and theory of singular integral equations can be extended to the case of differential forms and topological product of domains (Zhong Tongde [**Z1–6**]).

3. Holomorphic extension on Stein manifolds.

In this section we use Plemelj jump formula (1.16) on Stein manifold to discuss the problem of holomorphic extension of a function given values on the boundary ∂D into a relatively compact domain D in a Stein manifold M (Zhong Tongde [**Z7**], Lin Yaxian [**LinY**], C. Laurent-Thiebaut [**Lau**]). We have extended theorems of Hartogs and Bochner, which are well-known in $\mathbb{C}^n$ (A. Aizenberg and A. P. Yuzhakov [**AY**], F. R. Harvey and H. B. Lawson [**HLa**]).

Applying Plemelj jump formula (1.16) and complex Green formula we have proved the following

LEMMA 3.1. *Assume that D is a relatively compact domain on M with piecewise smooth boundary and the Bochner-Martinelli integral representation*

$$(3.1) \qquad \int_{\partial D} f(\zeta)\Omega^0(\phi^\nu, \bar{s}, s; z) = \begin{cases} f(z), & z \in D \\ 0, & z \notin \bar{D}, \end{cases}$$

is valid for $z \in D$ for a particular $f \in C^1(\bar{D})$. Then $f \in A(D)$.

From Lemma 3.1 we have immediately the following result.

THEOREM 3.1. *Suppose that the boundary ∂D of D is a smooth orientable manifold of class C^2, and $f \in C^1(\partial D)$. Then there exists a function $F \in C^1(\bar{D})$ holomorphic in D such that $F|_{\partial D} = f$, if and only if*

$$(3.2) \qquad \int_{\partial D} f(\zeta)\Omega^0(\phi^\nu, \bar{s}, s; z) = 0, \qquad z \in M\backslash\bar{D}$$

Furthermore, if $M\backslash\bar{D}$ is connected, then it suffices that (3.2) holds for $z \in M\backslash\tilde{D}$, $D \subset\subset \tilde{D} \subset\subset M$.

If condition (2.1) is satisfied, then the extension F can be expressed explicitly by

$$(3.3) \qquad F(z) = \int_{\partial D} f(\zeta)\Omega^0(\phi^\nu, \bar{s}, s; z), \qquad z \in D^+.$$

By Theorem 3.1 we can prove

THEOREM 3.2 (BOCHNER). *Suppose that the domain D has boundary of class C^2, $M\backslash\bar{D}$ is connected, and $f \in C^1(\partial D)$. Then there exists a function $F \in A(D) \cap C^1(\bar{D})$ such that $F|_{\partial D} = f$, if and only if*

$$(3.4) \qquad\qquad \bar{\partial}_b f = 0.$$

THEOREM 3.3 (HARTOGS, OSGOOD AND BRAUN). *Suppose that D is a relatively compact domain of M such that $M\backslash\bar{D}$ is connected. Then every function $f(z)$ holomorphic on ∂D (i.e., in some neighborhood $U(\partial D)$) can be extended holomorphically to the whole of D.*

4. $\bar{\partial}$-closed extension of (p, q) differential forms in $\mathbb{C}^n$.

It is well-known, the Bochner-Martinelli-Koppelman kernel in $\mathbb{C}^n \times \mathbb{C}^n$ assumes the form:

$$(4.1) \qquad K(z, \zeta) = \frac{(n-1)!}{(2\pi i)^n} \frac{\bar{\omega}'_{z,\zeta}(\bar{z} - \bar{\zeta}) \wedge \omega_{z,\zeta}(z - \zeta)}{|z - \zeta|^{2n}},$$

where

$$(4.2) \qquad \omega_{z,\zeta}(u) = \bigwedge_{j=1}^{n} d_{z,\zeta} u_j, \qquad \bar{\omega}'_{z,\zeta}(v) = \sum_{j=1}^{n} (-1)^{j-1} v_j \bigwedge_{k \neq j} \bar{\partial}_{z,\zeta} v_k.$$

Let D be a bounded domain in $\mathbb{C}^n$ such that its boundary ∂D is connected and piecewise continuous differentiable. Then for all (p, q) differential forms which are differentiable in the neighborhood $\bar{D}$, we have Koppelman formula

$$(-1)^{p+q} \left[\iint_{\zeta \in \partial D} f(\zeta) \wedge K_{p,q}(z, \zeta) - \int_{\zeta \in D} \bar{\partial}_\zeta f(\zeta) \wedge K_{p,q}(z, \zeta) \right.$$
$$\left. + \bar{\partial}_z \int_{\zeta \in D} f(\zeta) \wedge K_{p,q-1}(z, \zeta) \right] = \begin{cases} f(z), & z \in D^+, \\ 0, & z \in D^-, \end{cases}$$

where $K_{p,q}(z, \zeta)$ is the component of K of bidegree (p, q) in Z and $(n-p, n-q-1)$ in ζ.

DEFINITION 4.1. Given (p, q) differential form $f \in C^1_{p,q}(\partial D)$, then

$$(4.4) \qquad F^{\pm}(z) = \int_{\zeta \in \partial D} f(\zeta) \wedge K_{p,q}(z, \zeta), \qquad z \in D^{\pm}$$

is called the Bochner-Martinelli transform of f, $F^+(z)$ and $F^-(z)$ are complex harmonic (p, q) differential forms in D^+ and D^- respectively.

THEOREM 4.1 (JUMP FORMULA). *Let D be a bounded domain in $\mathbb{C}^n$ with its boundary ∂D being connected and piecewise continuous differentiable. If $f \in C^1_{p,q}(\bar{D})$, then for $F^{\pm}(z)$ we have jump formula*

$$(4.5) \qquad F^+(\eta) - F^-(\eta) = (-1)^{p+q} f(\eta), \qquad \eta \in \partial D.$$

Moreover, $F^\pm(z)$ can continuously be extended to the boundary.

Applying jump formula (4.5) and integral representation of Green type we have proved the following

LEMMA 4.1. *Let D be a bounded domain in $\mathbb{C}^n$ with its boundary being piecewise smooth. Assume $f \in C^1_{p,q}(\bar{D})$ and Koppelman formula*

$$(4.7) \quad f(z) = (-1)^{p+q} \left[\int_{\zeta \in \partial D} f(\zeta) \wedge K_{p,q}(z,\zeta) + \bar{\partial}_z \int_{\zeta \in D} f(\zeta) \wedge K_{p,q-1}(z,\zeta) \right]$$

holds. Then f is $\bar{\partial}$-closed.

As a corollary we have

THEOREM 4.2. *Suppose that the boundary ∂D of D is a smooth orientable manifold of class C^2, and $f \in C^1_{p,q}(\bar{D})$. Then there exists a form $G \in C^1_{p,q}(\bar{D})$ $\bar{\partial}$-closed in D with $G|_{\partial D} = f$, if and only if*

$$(4.8) \quad \int_{\zeta \in \partial D} f(\zeta) \wedge K_{p,q}(z,\zeta) + \bar{\partial}_z \int_{\zeta \in D} f(\zeta) \wedge K_{p,q-1}(z,\zeta) \equiv 0, \qquad z \in D^-.$$

If $\mathbb{C}^n \backslash \bar{D}$ is connected, then it suffices that (4.8) hold for $z \in \{ w \mid |w_i| > R, i = 1, \ldots, n \}$, where R is a sufficiently large positive number.

Moreover if condition (4.8) is satisfied, then the extension form for G can be expressed explicitly by

$$(4.9) \quad \begin{aligned} G(z) = (-1)^{p+1} & \left[\int_{\zeta \in \partial D} f(\zeta) \wedge K_{p,q}(z,\zeta) \right. \\ & \left. + \bar{\partial}_z \int_{\zeta \in D} f(\zeta) \wedge K_{p,q-1}(z,\zeta) \right], \qquad z \in \bar{D}. \end{aligned}$$

By Theorem 4.2 we can prove

THEOREM 4.3. *Assume D is a bounded domain with its boundary ∂D being connected such that $\mathbb{C}^n \backslash \bar{D}$ also is connected, $n > 1$. Then every (p,q) differential form $\bar{\partial}$-closed on ∂D (i.e., in some neighborhood $U(\partial D)$) can be extended to a $\bar{\partial}$-closed differential form on $\bar{D}$, that is, there exists a $\bar{\partial}$-closed form F on $\bar{D}$ such that $F|_{\partial D} = f$.*

The above results can also be extended to Stein manifolds (Zhong Tongde and Qiu Chunhui [**ZQ2**]).

Moreover, Zhong Tongde [**Z8**] obtained a $\bar{\partial}$-closed extension theorem of (p,q) differential forms in the sense of J. J. Kohn and A. V. Rossi [**KR**].

Let D be a relatively compact domain on Stein manifold M. We denote by $\mathcal{A}^{p,q}$ the space of C^∞ forms of type (p,q) on D, $\dot{\mathcal{A}}^{p,q}$ the space of (p,q)-forms which are C^∞ up to and including ∂D, i.e., consisting of restrictions of (p,q)-forms on M. Let $\tilde{\mathcal{A}}^{p,q}$ denote the space of (p,q)-forms on ∂D, $\tilde{\mathcal{D}}^{p,q}$ the subspace of $\tilde{\mathcal{A}}^{p,q}$ consisting of complex tangential forms. Let Λ denote the restriction map $\Lambda : \dot{A}^{p,q} \to \tilde{A}^{p,q}$ and $\tilde{\mu} : \tilde{A} \to \tilde{D}^{p,q}$ be a projection. We define the map $\mu : A^{p,q} \to \tilde{D}^{p,q}$ by $\mu = \tilde{\mu} \circ \Lambda$.

J. J. Kohn and A. V. Rossi [**KR**] proposed the following problem for the extension of $\bar{\partial}_b$-closed forms:

PROBLEM. Given $\phi \in \tilde{A}^{p,q}$ to find $\phi_0 \in \dot{A}^{p,q}$ such that $\bar{\partial}\phi_0 = 0$ and $\mu(\phi_0) = \tilde{\mu}(\phi)$. If such a ϕ_0 exists, then we say that it is a $\bar{\partial}$-closed extension of ϕ.

We have proved the following (Zhong Tongde [**Z8**]):

THEOREM 4.4. *If $D \subset\subset M$ is a strictly pseudoconvex domain on Stein manifold with ∂D being smoothly connected, then every $\bar{\partial}_b$-closed form $\phi \in \tilde{A}^{p,q}$ has $\bar{\partial}$-closed extension, which is a $\bar{\partial}$-exact extension.*

5. Plemelj formula in the case of piecewise smooth boundary.

Let D be a bounded domain in $\mathbb{C}^n$, its boundary ∂D is a piecewise closed smooth orientable manifold of class C^1 (R. Michael Range and Yum-Tong Siu [**RS**]), in the sense of geometrical meaning of Gauss integral. We have

DEFINITION 5.1. Assume $\eta \in \partial D$. If limit

$$(5.1) \qquad \lim_{\delta \to 0} \int_{\bar{D} \cap \partial S_\delta(\eta)} \frac{dS}{\delta^{2n-1}}$$

exists, then this limit is defined as the measure of volume angle from point η seeing over boundary ∂D, denoted by $G(\eta)$; set

$$(5.2) \qquad \sigma_n = \frac{2\pi^n}{(n-1)!}$$

(volume of unit hypersphere of $2n-1$ dimensions), then $\alpha(\eta) = \dfrac{G(\eta)}{\sigma_n}$ is said to be an angular coefficient of the point $\eta \in \partial D$, $0 \leqslant \alpha(\eta) \leqslant 1$. Similar to the case of one complex variable, when $\alpha(\eta) = \frac{1}{2}$, η is a smooth boundary point; $\alpha(\eta) = 0$ or 1, η is a outward cusp point on boundary or inward cusp point on boundary; $0 < \alpha(\eta) < 1$, and $\alpha(\eta) \neq \frac{1}{2}$, η is a generalized angular point of boundary.

We have (Lin Liangyu [**LinL**])

THEOREM 5.1. *Let ∂D be a closed piecewise smooth orientable manifold of class C^1, $\xi, \eta \in \partial D$. Then*

$$(5.3) \qquad \text{V.P.} \int_{\partial D} K(\xi, \eta) = \frac{G(\eta)}{\sigma_n} = \alpha(\eta), \qquad 0 < \alpha(\eta) < 1.$$

THEOREM 5.2. *If $f(\xi) \in H(\alpha, \partial D)$, $\eta \in \partial D$, then the Cauchy principal value of $F(\eta)$ in the sense of (1.4) exists.*

THEOREM 5.3. *If $f(\xi) \in H(\alpha, \partial D)$, $\eta \in \partial D$, then for the integral (1.3) we have Plemelj formula*

$$(5.4) \qquad F^+(\eta) = \text{V. P.} \int_{\partial D} f(\xi) K(\xi, \eta) + (1 - \alpha(\eta)) f(\eta),$$

$$(5.5) \qquad F^-(\eta) = \text{V. P.} \int_{\partial D} f(\xi) K(\xi, \eta) - \alpha(\eta) f(\eta),$$

$0 < \alpha(\eta) < 1.$

THEOREM 5.4. *If $f(\xi) \in H(\alpha, \partial D)$, $\eta \in \partial D$, then*

$$F^+(\eta) \text{ and } F^-(\eta) \in H(\alpha - \varepsilon, \partial D), \qquad 0 < \varepsilon < \alpha.$$

6. Finite part of singular integrals of Bochner-Martinelli type.

Assume that $0 \in \partial D$ and that in a neighborhood U of the origin. Then

$$(6.1) \qquad \partial D \cap U = \{z \in U \mid r(u_1, \ldots, u_{2n}) = 0\},$$

for a function $r \in C^3(U)$ satisfying

$$\left.\frac{\partial r}{\partial u_1}\right|_{u=0} = 1, \qquad \left.\frac{\partial r}{\partial u_2}\right|_{u=0} = \cdots = \left.\frac{\partial r}{\partial u_{2n}}\right|_{u=0} = 0.$$

In other words, the tangent plane of ∂D at the origin is defined by $u_1 = 0$. Since $\left.\frac{\partial r}{\partial u_1}\right|_{u=0} \neq 0$, we can write the equation $r(u_1, \ldots, u_{2n}) = 0$ as

$$(6.2) \qquad u_1 - h(u_2, \ldots, u_{2n}) = 0,$$

for a suitable function h of class C^3 and satisfying

$$h(0, \ldots 0) = 0, \qquad \left.\frac{\partial h}{\partial u_2}\right|_{u=0} = \cdots = \left.\frac{\partial h}{\partial u_{2n}}\right|_{u=0} = 0.$$

We have (Wang Xiaoqin [**WanX2**])

THEOREM 6.1. *Suppose $\frac{\partial h}{\partial u_i}$, $i = 2, \ldots, 2n$, are Hölder continuous on ∂D, and f is a continuously differentiable function on ∂D such that $\frac{\partial f}{\partial u_j}$, $j = 1, \ldots, 2n$, are Hölder continuous. Then the finite part*

$$(6.3) \qquad \text{F. P.} \int_{\partial D} f(\xi) \frac{\bar{\xi}_j}{|\xi|^2} K(\xi, 0)$$

defined as

$$(6.4) \qquad \lim_{\varepsilon \to 0} \left[\int_{\partial D \setminus S_\varepsilon(0)} f(\xi) \frac{\bar{\xi}_j}{|\xi|^2} K(\xi, 0) - \frac{B_j f(0)}{\varepsilon} \right],$$

exists, where

$$(6.5) \qquad B_j = \begin{cases} \dfrac{(-2)^{n-1}(n-1)!}{\pi(2n-1)!!}, & j = 1, \\[2mm] 0, & j = 2, \ldots, n. \end{cases}$$

The following extension of Plemelj formula holds.

THEOREM 6.2. *Let f and D be as in Theorem 6.1. Then for the integral* (1.3), *the limits of* $\left(\dfrac{\partial F}{\partial z_j}\right)(z)$, $j = 1, \ldots n$, *exist as z tends to 0 along the inner normal and the exterior normal of ∂D at the point 0, and are given by the following expressions:*

(6.6)
$$\left(\frac{\partial F}{\partial z_j}\right)_i (0) = \text{F.P.} \int_{\partial D} n f(\xi) \frac{\bar{\xi}_j}{|\xi|^2} K(\xi, 0)$$
$$+ \frac{1}{2}\frac{\partial f}{\partial \xi_j}(0) + \frac{1}{2} A_j \frac{\partial f}{\partial \bar{\xi}_j}(0);$$

(6.7)
$$\left(\frac{\partial F}{\partial z_j}\right)_e (0) = \text{F.P.} \int_{\partial D} n f(\xi) \frac{\bar{\xi}_j}{|\xi|^2} K(\xi, 0)$$
$$- \frac{1}{2}\frac{\partial f}{\partial \xi_j}(0) - \frac{1}{2} A_j \frac{\partial f}{\partial \bar{\xi}_j}(0),$$

where

(6.8)
$$A_j = \begin{cases} (-1)^n, & j = 1, \\ 0, & j = 2, \ldots, n. \end{cases}$$

Because $K(\xi, z)$ does not change under a unitary transformation, the finite part of f in ∂D at point η equals that of $g(z) = f\left(\sum_{j=1}^n U_{jk} z_j\right)$ in ∂D^* at the point $z = 0$, where (U_{jk}) is a unitary matrix, that is the above results is also hold in the general case.

II. Integral representations in several complex variables

1. An integral formula on a bounded domain in $\mathbb{C}^n$.

Let D be a bounded domain with piecewise smooth boundary in $\mathbb{C}^n$, $\lambda_2, \lambda_3,$ $\ldots, \lambda_m$ ($m < +\infty$) are $m-1$ independent arbitrary real parameters in $\mathbb{R}$, vector function $w = (w_1, \ldots, w_n) \in C^1(\partial D)$ and $\langle w(\zeta, z), \zeta - z \rangle \neq 0$, for $\zeta \in \partial D$, $z \in D$.
Denote

(1.1)
$$g_\alpha^{(m)} = \frac{(\bar{\zeta}_\alpha - \bar{z}_\alpha)|\zeta_\alpha - z_\alpha|^{m-2}}{\sum_{j=1}^n |\zeta - z|^m},$$

$\zeta \in \partial D$, $z \in D$, $\alpha = 1, 2, \ldots, n$,

(1.2)
$$h_\alpha = \frac{w_\alpha}{\langle \zeta - z, w \rangle},$$

$w = (w_1, \ldots, w_n)$, $\zeta \in \partial D$, $z \in D$, $\alpha = 1, 2, \ldots, n$, where $m = 2, 3, \ldots, N$,

$(N < +\infty)$. Now define a recursive operator $T^{(m)}$ as following:

$$T^{(m)}h_\alpha = \lambda_m g_\alpha^{(m)} + (1 - \lambda_m)T^{(m-1)}h_\alpha,$$
$$T^{(m-1)}h_\alpha = \lambda_{m-1}g_\alpha^{(m-1)} + (1 - \lambda_{m-1})T^{(m-2)}h_\alpha,$$

(1.3)
$$\ldots\ldots$$

$$T^{(2)}h_\alpha = \lambda_2 g_\alpha^{(2)} + (1 - \lambda_2)T^{(1)}h_\alpha,$$
$$T^{(1)}h_\alpha = h_\alpha.$$

Obviously the denominator in the vector operator $T^{(m)}h_\alpha$ not equal to zero on boundary ∂D.

By induction we can prove

LEMMA 1.1. *Let*

(1.4) $H_\alpha^{(m)} = T^{(m)}h_\alpha, \qquad m = 2, 3, \ldots, N, (N < +\infty),$

(1.5) $H_\alpha^{(1)} = T^{(1)}h_\alpha = h_\alpha, \qquad \alpha = 1, 2, \ldots, n.$

Then $H_\alpha^{(m)}(\lambda_2, \lambda_3, \ldots, \lambda_m, \zeta, z, w)$ *is a partition of unit of bounded domain* D.

Using this partition of unit we can define a kernel of bounded domain D as following:

$$\Omega(\lambda_2, \ldots, \lambda_m, \zeta, z, w) = \frac{(n-1)!}{(2\pi i)^n}\left(\sum_{\alpha=1}^{n}(-1)^{\alpha-1}H_\alpha^{(m)}dH_1^{(m)}\wedge\right.$$

(1.6)
$$\left.\wedge\cdots\wedge[\alpha]\wedge\cdots\wedge dH_n^{(m)}\right)\wedge d\zeta,$$

$(-\infty < \lambda_2, \lambda_3, \ldots, \lambda_m < +\infty)$, where $m = 2, 3, \ldots, N, (N < +\infty)$, and

(1.7)
$$H_\alpha^{(m)} = T^{(m)}h_\alpha = \lambda_m g_\alpha^{(m)} + (1 - \lambda_m)T^{(m-1)}h_\alpha$$
$$= \lambda_m g_\alpha^{(m)} + (1 - \lambda_m)H_\alpha^{(m-1)}.$$

Moreover $d = d\lambda_m + d\zeta = d\lambda_m + \partial\zeta + \bar{\partial}\zeta$, $\lambda_2, \lambda_3, \ldots, \lambda_{m-1}$ are arbitrary independent real parameters and are independent of λ_m.

For an arbitrary $z \in D$, consider the domain:

(1.8) $G = \{\lambda_m \in [0, 1], \zeta \in \partial D; \lambda_m = 1, D\backslash B_\varepsilon(z)\}$

where $B_\varepsilon(z)$ is a hypersphere with radius $\varepsilon > 0$, with center at z, the boundary of G is:

$$\partial G = \{\lambda_m = 0, \zeta \in \partial D; \lambda_m = 1, \zeta \in \partial B_\varepsilon(z)\}.$$

Since $d\Omega(\lambda_2, \ldots, \lambda_m, \zeta, z, w) = 0$ on D, then for any $f(z) \in C^\infty(\bar{D})$, we have

(1.9)
$$d[f(\zeta)\Omega(\lambda_2, \ldots, \lambda_{m-1}, \lambda_m, \zeta, z, w)]$$
$$= \bar{\partial}f(\zeta) \wedge \Omega(\lambda_2, \ldots, \lambda_{m-1}, \lambda_m, \zeta, z, w).$$

Applying Stokes formula on G, we have (Yao Zongyuan [Y])

THEOREM 1.1. *Let D be a bounded domain with piecewise smooth boundary ∂D in $\mathbb{C}^n$, $\lambda_2, \lambda_3, \ldots, \lambda_m$, $(m < +\infty)$ are $m - 1$ arbitrary independent real parameters in $\mathbb{R}$. If vector function $w \in C^1(\partial D)$ and $\langle w(\zeta, z), \zeta - z \rangle \neq 0$, where $\zeta \in \partial D$, $z \in D$, then any function $f(z) \in C^\infty(\bar{D})$ can be represented by*

$$
\begin{aligned}
f(z) = &\int_{\partial D} f(\zeta) \Omega(\lambda_2, \ldots, \lambda_{m-1}, 0, \zeta, z, w) \\
&- \int_D \bar{\partial}_\zeta f(\zeta) \omega^{(m)}(\zeta - z, \bar{\zeta} - \bar{z}) \\
&+ \int_{\partial D} \bar{\partial}_\zeta f(\zeta) \wedge \int_{\lambda_m \in [0,1]} \Omega(\lambda_2, \ldots, \lambda_{m-1}, \lambda_m, \zeta, z, w),
\end{aligned}
\tag{1.10}
$$

$z \in D$, $-\infty < \lambda_2, \ldots, \lambda_{m-1} < +\infty$, $m = 2, 3, \ldots, N, (N < +\infty)$, *where*

$$
\begin{aligned}
\omega^{(m)}(\zeta - z, \bar{\zeta} - \bar{z}) = &\frac{(n-1)!}{(2\pi i)^n} \left(\frac{m}{2}\right)^{n-1} \prod_{j=1}^n |\zeta_j - z_j|^{m-2} \\
&\frac{\sum_{\alpha=1}^n (-1)^{\alpha-1}(\bar{\zeta}_\alpha - \bar{z}_\alpha) d\bar{\zeta}_{[\alpha]} \wedge d\zeta}{\left[\sum_{j=1}^n |\zeta_j - z_j|^m\right]^n},
\end{aligned}
\tag{1.11}
$$

$$
\begin{aligned}
&\Omega(\lambda_2, \ldots, \lambda_{m-1}, \lambda_m, \zeta, z, w) \\
&= \frac{(n-1)!}{(2\pi i)^n} \left(\sum_{\alpha=1}^n (-1)^{\alpha-1} H_\alpha^{(m)} dH_1^{(m)}\right. \\
&\qquad\left. \wedge \cdots \wedge [\alpha] \wedge \cdots \wedge dH_n^{(m)}\right) \wedge d\zeta,
\end{aligned}
\tag{1.12}
$$

$$
\begin{aligned}
&\Omega(\lambda_2, \ldots, \lambda_{m-1}, 0, \zeta, z, w) \\
&= \frac{(n-1)!}{(2\pi i)^n} \left(\sum_{\alpha=1}^n (-1)^{\alpha-1} H_\alpha^{(m-1)} dH_1^{(m-1)}\right. \\
&\qquad\left. \wedge \cdots \wedge [\alpha] \wedge \cdots \wedge dH_\alpha^{(m-1)}\right) \wedge d\zeta.
\end{aligned}
\tag{1.13}
$$

Especially, when $f(z)$ is holomorphic in D, then $\bar{\partial}_\zeta f(\zeta) = 0$, and we get a extension form of well-known Cauchy-Fantappie formula for holomorphic functions.

THEOREM 1.2. *Let D be a bounded domain with piecewise smooth boundary ∂D, $\lambda_2, \lambda_3, \ldots, \lambda_p$, $(p < +\infty)$ are $p - 1$ arbitrary independent real parameters in $\mathbb{R}$. If vector function $w \in C^1(\partial D)$ and $\langle w(\zeta, z), \zeta - z \rangle \neq 0$, where $\zeta \in \partial D$, $z \in D$, then any function $f(z)$ which is holomorphic in D continuous on $\bar{D}$, can be represented by*

$$
f(z) = \int_{\partial D} f(\zeta) K(\lambda_2, \lambda_3, \ldots, \lambda_p, \zeta, z, w),
\tag{1.14}
$$

$z \in D$, $p = 2, 3, \ldots, N$, $(N < +\infty)$, where

(1.15)
$$K(\lambda_2, \lambda_3, \ldots, \lambda_p, \zeta, z, w) = \frac{(n-1)!}{(2\pi i)^n} \left(\sum_{\alpha=1}^n (-1)^{\alpha-1} H_\alpha^{(p)} dH_1^{(p)} \right.$$
$$\left. \wedge \cdots \wedge [\alpha] \wedge \cdots \wedge dH_n^{(p)} \right) \wedge d\zeta.$$

Especially, if $\lambda_2 = \lambda_3 = \cdots = \lambda_p = 0$, then

(1.16)
$$K(\lambda_2, \ldots, \lambda_p, \zeta, z, w) = \omega(\zeta - z, w),$$

and (1.14) is the well-known Cauchy-Fantappie formula

(1.17)
$$f(z) = \frac{(n-1)!}{(2\pi i)^n} \int_{\partial D} f(\zeta) \frac{\sum_{\alpha=1}^n (-1)^{\alpha-1} w_\alpha dw_{[\alpha]} \wedge d\zeta}{\langle \zeta - z, w \rangle^n}.$$

As usually from formulas (1.10) and (1.14) we can deduce some extension forms of well-known formulas, such as Cauchy-Fantappie formula, Bochner-Martinelli formula, Leray-Stokes formula, Henkin-Ramirez formula and so forth.

2. Integral representation of (p, q) differential forms on real non-degenerate strictly pseudoconvex polyhedra on Stein manifolds.

2.1 Koppelman-Leray-Norguet formula.

Let M be a Stein manifold, and n be the complex dimension of M. If $D \subset\subset M$ is an open set, the boundary ∂D of D is of piecewise C^1. Set $S_k := \{z \in \partial D \cap V_k \mid \rho_k(z) = 0\}$ for $k = 1, \ldots, N$, where V_k and ρ_k are as usually in the definition of a piecewise C^1-boundary. For every ordered collection $K = (k_1, \ldots, k_\ell)$ of integers $1 \leqslant k_1, \ldots, k_\ell \leqslant N$, we define (G. M. Henkin and J. Leiterer [**HLe**])

$$S_K := \begin{cases} S_{k_1} \cap \cdots \cap S_{k_\ell}, & \text{if the integers } k_1, \ldots, k_\ell \text{ are different in pairs,} \\ \emptyset, & \text{otherwise.} \end{cases}$$

We choose the orientation on S_K so that

$$\partial D = \sum_{k=1}^N S_k$$

and

$$\partial S_K = \sum_{j=1}^N S_{K_j},$$

where $K_j := (k_1, \ldots, k_\ell, j)$ if $K = (k_1, \ldots, k_\ell)$.

We denote by Δ the subset of all points $(\lambda_0, \lambda_1, \ldots, \lambda_N) \in \mathbb{R}^{N+1}$ such that $\lambda_j \geqslant 0$ for $j = 0, 1, \ldots, N$ and $\sum_{j=0}^N \lambda_j = 1$. For every strictly increasing collection $I = (i_1, \ldots, i_\ell)$ of integers $0 \leqslant i_1 < \cdots < i_\ell \leqslant N$, we set

$$\Delta_I := \left\{ \lambda \in \Delta \,\middle|\, \sum_{r=1}^\ell \lambda_{i_r} = 1 \right\},$$

$OK := (0, k_1, \ldots, k_\ell)$ for every collection $I = (i_1, \ldots, i_\ell)$ of integers we write $|I| := \ell$.

Denote tangent bundle and cotangent bundle of Stein manifold M by $T(M)$ and $T^*(M)$ respectively, $\tilde{T}(M \times M)$ and $\tilde{T}^*(M \times M)$ are the pull-back with respect to the projection $(z, \zeta) \to z$. Denote $\bar{T}^*(M \times M \times \Lambda)$ as the vector bundle of $M \times M \times \Lambda$, which is the pull-back of $T^*(M)$ with respect to the projection $(z, \zeta, \lambda) \to z$.

Suppose that θ is a C^∞ Hermite metric, it induces a Hermite metric on $\tilde{T}(M \times M)$, also denote by θ, and it induces a metric θ^* on $\tilde{T}^*(M \times M)$. Assume that D is a connection on $\tilde{T}(M \times M)$ admissible for θ, ∇ is a connection on $\tilde{T}^*(M \times M)$ admissible for θ^*.

Let $(s_1^*, \ldots, s_N^*, k^*)$ be a Leray-Norguet section of (D, s, ϕ). Denote

$$(2.1) \qquad t^*(z, \zeta, \lambda) = \sum_{k \in K} \lambda_k \frac{s_k^*(z, \zeta)}{\langle s_k^*(z, \zeta), s(z, \zeta) \rangle} + \lambda_0 \frac{\hat{s}(z, \zeta)}{\|s(z, \zeta)\|_0^2},$$

where $s_k^*(z, \zeta) \in T_z^*(M)$, for all $z \in D$, ζ belongs to a neighborhood of ∂D, $\hat{s} = \sigma s$, $s(z, \zeta) \in T_z(M)$, σ is a anti-linear map, $\sigma : \tilde{T}(M \times M) \to \tilde{T}^*(M \times M)$.

Define

$$(2.2) \qquad \bar{\Omega}(\phi^\nu, s_1^*, \ldots, s_N^*, \hat{s}, s) = \frac{(-1)^{n-1}}{(2\pi)^n} \phi^{\nu n} \langle t^*, DS \rangle \wedge (\langle \Delta'' t^*, DS \rangle)^{n-1},$$

where $\nu \geqslant \max(k, k^*)$, which is continuous in a neighborhood of $D \times \partial D \times \Lambda$. Applying Koppelman formula for (p, q) differential form (J. P. Demailly and C. Laurent-Thiebaut [DL], see also Zhong Tongde and Huang Sha [ZH]) we can prove (Wang Zhiqiang [WanZ])

THEOREM 2.1 (KOPPELMAN-LERAY-NORGUET). *Let D be a relative compact set on a Stein manifold, its boundary ∂D be piecewise smooth of C^1, $(s_1^*, \ldots, s_N^*, k^*)$ be a Leray-Norguet section of (D, s, ϕ), $\nu \geqslant \max(2k, k^*)$ is integer. Assume that for $\phi^\nu s_k^* / \langle s_k^*, s \rangle$, all derivatives with respect to z of degree $\leqslant 2$, and all derivatives with respect to ζ of degree $\leqslant 1$ are continuous in a neighborhood $W_k \subset D \times M$ of $D \times S_k$. Suppose that f is a continuous (p, q) form on $\bar{D}$, $\bar{\partial} f$ is also continuous on $\bar{D}$. Then for set $K = (k_1, \ldots, k_\ell)$ which consisted by all strictly increase integers $1 \leqslant k_1 < \cdots < k_\ell \leqslant N$, we have*

$$(2.3) \qquad (-1)^{p+q} f(z) = \sum_{|K| \leqslant n-p-q} \int_{(\zeta, \lambda) \in S_K \times \Lambda_K} f(\zeta)$$

$$\wedge \bar{\Omega}_q^p(\phi^\nu, s_1^*, \ldots, s_N^*, s)$$

$$- \int_{\zeta \in D} \bar{\partial}_\zeta f(\zeta) \wedge \tilde{\Omega}_q^p(\phi^\nu, \hat{s}, s)$$

$$- \sum_{|K| \leqslant n-p-q-1} (-1)^{|K|} \left(\int_{(\zeta, \lambda) \in S_K \times \Lambda_{OK}} \bar{\partial}_\zeta f(\zeta) \right.$$

$$\wedge \bar{\Omega}_q^p(\phi^\nu, s_1^*, \ldots, s_N^*, \hat{s}, s)\Big)$$

$$+ \sum_{|K| \leqslant n-p-q} (-1)^{|K|} \Big(\bar{\partial}_z \int_{(\zeta,\lambda) \in S_K \times \Lambda_{OK}} f(\zeta)$$

$$\wedge \bar{\Omega}_{q-1}^p(\phi^\nu, s_1^*, \ldots, s_N^*, \hat{s}, s)$$

$$+ (-1)^{p+q+1} \int_{(\zeta,\lambda) \in S_K \times \Lambda_{OK}} f(\zeta)$$

$$\wedge Q_q^p(\phi^\nu, s_1^*, \ldots, s_N^*, \hat{s}, s)\Big)$$

$$+ \bar{\partial}_z \int_{\zeta \in D} f(\zeta) \wedge \tilde{\Omega}_{q-1}^p(\phi^\nu, \hat{s}, s)$$

$$+ (-1)^{p+q+1} \int_{\zeta \in D} f(\zeta) \wedge P_q^p(\phi^\nu, \hat{s}, s).$$

2.2 Formula of (p,q)-forms on real non-degenerate strictly pseudoconvex poly-hedra.

Assume that D is a real non-degenerate strictly pseudoconvex polyhedra on a Stein manifold M (definition see G. M. Henkin and J. Leiterer [**HLe**]). Let $(h_1^*, \ldots, h_N^*, 1)$ be a Leray-Norgeut section of (D, s, ϕ), which is holomorphic with respect to $z \in D$ (whose existence see G. M. Henkin and J. Leiterer [**HLe**]), from Theorem 2.1 we have immediately (Wang Zhiqiang [**WanZ**])

THEOREM 2.2. *Assume that D is a real non-degenerate strictly pseudoconvex polyhedra on a Stein manifold M, $\nu \geqslant 2k$ is integer, and $c(\tilde{T}(M \times M)) = 0$, $q \geqslant 1$.*

i) *If f is a (p,q)-form continuous on $\bar{D}$, $\bar{\partial}f$ also continuous on $\bar{D}$, then*

$$f(z) = (-1)^{p+q} \Big[\bar{\partial}_z \int_{\zeta \in D} f(\zeta) \wedge \tilde{\Omega}_{q-1}^p(\phi^\nu, \hat{s}, s)$$

$$- \int_{\zeta \in D} \bar{\partial}_\zeta f(\zeta) \wedge \tilde{\Omega}_q^p(\phi^\nu, \hat{s}, s)$$

(2.4)
$$+ \sum_{|K| \leqslant n-p-q} (-1)^{|K|} \bar{\partial}_z \int_{(\zeta,\lambda) \in S_K \times \Lambda_{OK}} f(\zeta)$$

$$\wedge \bar{\Omega}_{q-1}^p(\phi^\nu, h_1^*, \ldots, h_N^*, \hat{s}, s)$$

$$- \sum_{|K| \leqslant n-p-q-1} (-1)^{|K|} \int_{(\zeta,\lambda) \in S_K \times \Lambda_{OK}} \bar{\partial}_\zeta f(\zeta)$$

$$\wedge \bar{\Omega}_q^p(\phi^\nu, h_1^*, \ldots, h_N^*, \hat{s}, s)\Big]$$

ii) *If f is a (p,q)-form continuous on $\bar{D}$, such that $\bar{\partial}f = 0$, then*

$$g(z) = (-1)^{p+q} \left(\int_{\zeta \in D} f(\zeta) \wedge \tilde{\Omega}^p_{q-1}(\phi^\nu, \hat{s}, s) \right.$$

(2.5)
$$+ \sum_{|K| \leqslant n-p-q} (-1)^{|K|} \int_{(\zeta, \lambda) \in S_K \times \Lambda_{OK}} f(\zeta)$$

$$\left. \wedge \bar{\Omega}^p_{q-1}(\phi^\nu, h_1^*, \ldots, h_N^*, \hat{s}, s) \right)$$

is a continuous solution of $\bar{\partial}$-equation $\bar{\partial}g = f$.

3. Integral representation of (p,q) differential forms with weight factors on complex manifolds.

Consider a general complex manifold X (e.g. $M \times M$, M complex manifold) and a complex submanifold Y of X (e.g. $Y = \Delta$, Δ is the diagonal in $M \times M$). Suppose that we have given a complex vector-bundle $E \xrightarrow{\pi} X$ of rank $p = \text{codim}\, Y$, and let η be a smooth section to E, and $Y = \{\eta = 0\}$, moreover

$$d\eta_1 \wedge \cdots \wedge d\eta_p \wedge d\bar{\eta}_1 \wedge \cdots \wedge d\bar{\eta}_p \neq 0 \text{ on } Y$$

where the η_j:s are the coefficients of η with respect to some frame.

Let E^* be the dual bundle of E, ξ be an arbitrary section of E^*, D and D^* are the connections of E and E^* (D^* is the dual of D). If $e_1, \ldots, e_p$ is a local frame for E and $e_1^*, \ldots, e_p^*$ the dual frame for E^*, then $e_1^* \wedge e_1 \wedge \cdots \wedge e_p^* \wedge e_p = \Lambda$ is independent of the frames.

Let $c_p[\Theta]$ be the pth Chern form of Θ, Θ be the curvature form and $\Theta = D^2$. Let $\Theta = (\theta_{ij})_{p \times p}$, define

(3.1)
$$\tilde{\Theta} = \sum \theta_{ij} e_i^* \wedge e_j.$$

Then $\tilde{\Theta}$ is independent of local frame, and we have

(3.2)
$$\left(\frac{i}{2\pi} \right)^p \frac{1}{p!} \tilde{\Theta}^p = c_p[\Theta] e_1^* \wedge e_1 \wedge \cdots \wedge e_p^* \wedge e_p.$$

Now we construct a Cauchy-Leray kernel K with weight factors (M. Andersson [A]). Let $Q_0, \ldots, Q_N$ be smooth sections of E^*. Then $\langle Q_j, \eta \rangle$ are smooth functions on $X \times X$, and

(3.3)
$$d\langle Q_j, \eta \rangle = \langle D^* Q_j, \eta \rangle + \langle Q_j, D\eta \rangle$$

is a 1-form on $X \times X$. Set

(3.4)
$$\mathcal{M}\Lambda = C_p \sum_{k=0}^{p-1} \binom{p}{k} (-1)^{k-1} \sum_{|\alpha|=p-k} \frac{1}{\alpha!} g^{(\alpha)} (D^* Q_0 \wedge D\eta)^{\alpha_0}$$

$$\wedge \cdots \wedge (D^* Q_N \wedge D\eta)^{\alpha_N} \wedge \tilde{\Theta}^k,$$

where $C_p = \dfrac{1}{(2\pi i)^p p!}$, $\alpha_0 + \cdots + \alpha_N = p - k$, $\alpha! = \alpha_0! \ldots \alpha_N!$, $g(\lambda_0, \ldots, \lambda_N)$ is a holomorphic function, and

$$(3.5) \qquad g^{(\alpha)} = \left[\frac{\partial^{\alpha_0}}{\partial \lambda_0^{\alpha_0}} \cdots \frac{\partial^{\alpha_N}}{\partial \lambda_N^{\alpha_N}} \right] g(\langle Q_0, \eta \rangle, \ldots, \langle Q_N, \eta \rangle).$$

Let $\mathcal{N} = d\mathcal{M}$. Then $\mathcal{N}\Lambda = d\mathcal{M}\Lambda = (d\mathcal{M})\Lambda$. Now we replace g in $\mathcal{M}$ and $\mathcal{N}$ by $H(\lambda_0)g(\lambda_1, \ldots, \lambda_N)$ and Q_0 by $t\xi$, where t is a real parameter. We denote by $\mathcal{M}_t$ and $\mathcal{N}_t$ the forms so obtained when differentials also are taken with respect to t.

Denote

$$(3.6) \qquad (D^* Q \wedge D\eta)^\alpha = (D^* Q_1 \wedge D\eta)^{\alpha_1} \wedge \cdots \wedge (D^* Q_N \wedge D\eta)^{\alpha_N}.$$

Thus

$$(3.7) \qquad \begin{aligned} \mathcal{M}_t \Lambda &= C_p \sum_{k=0}^{p-1} \binom{p}{k} (-1)^{k-1} \sum_{j+|\alpha|=p-k} \frac{1}{j! \alpha!} H^{(j)} g^{(\alpha)} \\ &\quad (dt\xi \wedge D\eta + t D^*\xi \wedge D\eta)^j \wedge (D^* Q \wedge D\eta)^\alpha \wedge \tilde{\Theta}^k, \end{aligned}$$

$$(3.8) \qquad \begin{aligned} \mathcal{N}_t \Lambda &= C_p \sum_{k=0}^{p-1} \binom{p}{k} (-1)^{k-1} \sum_{j+|\alpha|=p-k} \frac{1}{j! \alpha!} \\ &\quad \{ dH^{(j)} \wedge g^{(\alpha)} (dt\xi \wedge D\eta + t D^*\xi \wedge D\eta)^j \wedge (D^* Q \wedge D\eta)^\alpha \wedge \tilde{\Theta}^k \\ &\quad + H^{(j)} dg^{(\alpha)} \wedge (dt\xi \wedge D\eta + t D^*\xi \wedge D\eta)^j \wedge (D^* Q \wedge D\eta)^\alpha \wedge \tilde{\Theta}^k \\ &\quad + H^{(j)} g^{(\alpha)} d(dt\xi \wedge D\eta + t D^*\xi \wedge D\eta)^j \wedge (D^* Q \wedge D\eta)^\alpha \wedge \tilde{\Theta}^k \\ &\quad + H^{(j)} g^{(\alpha)} (dt\xi \wedge D\eta + t D^*\xi \wedge D\eta)^j \wedge d[(D^* Q \wedge D\eta)^\alpha \wedge \tilde{\Theta}^k] \} \end{aligned}$$

PROPOSITION 3.1. *If $\mathcal{M}_t$ and $\mathcal{N}_t$ are as above, then*

$$(3.9) \qquad (d_t + d)\mathcal{M}_t = \mathcal{N}_t$$

where d is a differential does not contain t.

DEFINITION 3.1. Set $K = \displaystyle\int_0^\infty \mathcal{M}_t$, K is said as Cauchy-Leray kernel with factors.

Set $R = \displaystyle\int_0^\infty \mathcal{N}_t$, then

$$dK = \int_0^\infty d\mathcal{M}_t = -\int_0^\infty d_t \mathcal{M}_t + \int_0^\infty \mathcal{N}_t = \mathcal{M}_t|_{t=0} + R = P + R,$$

where we assume $\langle \xi, \eta \rangle \neq 0$.

The explicit expression of K, P, R are as following:

$$(3.10) \quad K\Lambda = \int_0^\infty \mathcal{M}_t \Lambda = C_p \sum_{k=0}^{p-1} \binom{p}{k}(-1)^k \sum_{j+|\alpha|=p-k} \frac{1}{\alpha!} g^{(\alpha)}$$

$$\frac{\xi \wedge D\eta \wedge (D^*Q \wedge D\eta)^{j-1} \wedge (D^*Q \wedge D\eta)^\alpha}{\langle \xi, \eta \rangle^j} \wedge \tilde{\Theta}^k,$$

$$(3.11) \quad \begin{aligned} P\Lambda &= \mathcal{M}_t|_{t=0}\Lambda \\ &= C_p \sum_{k=0}^{p-1} \binom{p}{k}(-1)^{k-1} \sum_{|\alpha|=p-k} \frac{1}{\alpha!} g^{(\alpha)}(D^*Q \wedge D\eta)^\alpha \wedge \tilde{\Theta}^k, \end{aligned}$$

(3.12)

$$R\Lambda = \int_0^\infty \mathcal{N}_t \Lambda = C_p \sum_{k=0}^{p-1} \binom{p}{k}(-1)^{k-1} \sum_{j+|\alpha|=p-k} \frac{1}{\alpha!}$$

$$\left\{ g^{(\alpha)} \left[\frac{j(\langle D^*\xi, \eta \rangle + \langle \xi, D\eta \rangle) \wedge \xi \wedge D\eta \wedge (D^*\xi \wedge D\eta)^{j-1}}{\langle \xi, \eta \rangle^{j+1}} \right. \right.$$

$$\left. - \frac{\langle \xi, \eta \rangle (D^*\xi \wedge D\eta)^j}{\langle \xi, \eta \rangle^{j+1}} \right] \wedge (D^*Q \wedge D\eta)^\alpha \wedge \tilde{\Theta}^k$$

$$- \sum_{\ell=1}^N \frac{\partial}{\partial \lambda_\ell} g^{(\alpha)}(\langle D^*Q_\ell, \eta \rangle + \langle Q_\ell, D\eta \rangle)$$

$$\wedge \frac{\xi \wedge D\eta \wedge (D^*\xi \wedge D\eta)^{j-1}}{\langle \xi, \eta \rangle^j} \wedge (D^*Q \wedge D\eta)^\alpha \wedge \tilde{\Theta}^k + g^{(\alpha)}$$

$$\left[\frac{(D^*\xi \wedge D\eta)^j - (d\xi \wedge D\eta + \xi \wedge C^*\xi \wedge D\eta - D^*\xi \wedge C\eta)(D^*\xi \wedge D\eta)^{j-1}}{\langle \xi, \eta \rangle^j} \right.$$

$$\left. + \frac{(j-1)\xi \wedge D\eta \wedge (C^*\xi \wedge D\eta - D^*\xi \wedge C\eta) \wedge (D^*\xi \wedge D\eta)^{j-2}}{\langle \xi, \eta \rangle^j} \right]$$

$$\wedge (D^*Q \wedge D\eta)^\alpha \wedge \tilde{\Theta}^k$$

$$+ g^{(\alpha)} \frac{\xi \wedge D\eta \wedge (D^*\xi \wedge D\eta)^{j-1}}{\langle \xi, \eta \rangle^j} \wedge \left[\sum_{\ell=1}^N \alpha_\ell (C^*Q_\ell \wedge D\eta - D^*Q_\ell \wedge C\eta) \right.$$

$$\left. \left. (D^*Q \wedge D\eta)^{\alpha_1,\ldots,\alpha_\ell-1,\ldots,\alpha_N} \wedge \tilde{\Theta}^k + (D^*Q \wedge D\eta)^\alpha \wedge k d\tilde{\Theta} \wedge \tilde{\Theta}^k \right] \right\}.$$

Especially, when $g \equiv 1$, $Q_\ell \equiv 0$, $\ell = 1, \ldots, N$, then we have

$$(3.13) \quad K\Lambda = C_p \sum_{k=0}^{p-1} \binom{p}{k}(-1)^k \frac{\xi \wedge D\eta \wedge (D\xi \wedge D\eta)^{p-k-1}}{\langle \xi, \eta \rangle^{p-k}} \wedge \tilde{\Theta}^k,$$

which is the Cauchy-Leray kernel constructed by B. Berndtson [B].

Let M be an n dimensional complex manifold, $X = M \times M$, E be a holomorphic vector bundle of rank n on X, η be a holomorphic section of E, such that

(3.14) $\Delta = \{\eta = 0\} = \{(\zeta, z) \in M \times M \mid \zeta = z\}.$

DEFINITION 3.2. Smooth section ξ of E^* is said to admissible for η, if for any compact set $K \subset X$,

(3.15) $|\xi| \leqslant C_K |\eta| \qquad \text{and} \quad |\langle \xi, \eta \rangle| \geqslant C_K |\eta|^2.$

Thus we have (Wu Xiaoqin [**Wu**])

THEOREM 3.1.
$$dK = [\Delta] + R + P$$
where ξ is admissible for η.

THEOREM 3.2. *Let D be a smoothly bounded domain in M, f be a smooth form of degree k on $\bar{D}$. Then for $z \in D$*

(3.16) $f(z) = \displaystyle\int_{\partial D} K \wedge f + \int_D K \wedge df + d \int_D K \wedge f + \int_D R \wedge f + \int_D P \wedge f.$

If E is a holomorphic bundle and η and D are holomorphic, then for a smooth (p, q) form f on $\bar{D}$, we have

(3.17)
$$f(z) = \int_{\partial D} K_{p,q} \wedge f + \int_D K_{p,q} \wedge \bar{\partial} f$$
$$+ \bar{\partial} \int_D K_{p,q-1} \wedge f + \int_D R_{p,q} \wedge f + \int_D P_{p,q} \wedge f,$$

where $K_{p,q}$, $R_{p,q}$, $P_{p,q}$ denotes the component of K, R, P of bidegree (p, q) in z respectively.

4. Fundamental solutions with weight factors of Cauchy-Riemann and induced Cauchy-Riemann operators on Stein manifolds.

Wang Xiaoqin [**WanX1**] using the result of B. Berndtson and M. Andersson [**BA**] extended the result of R. Harvey and J.Polking [**HP**] about the fundamental solutions of Cauchy-Riemann and induced Cauchy-Riemann operators to the case of with weight factors. Qiu Chunhui [**Q**] combined the methods of B. Berndtson and M. Andersson [**BA**] and J. P. Demailly and C. Laurent-Thiebaut [**DL**] extended the result of R. Harvey and J. Polking [**HP**] and Wang Xiaoqin [**WanX1**] to the Stein manifolds.

Firstly, by using Chern metric and connection, the Bochner-Martinelli kernel $B(z, \zeta)$, Leray kernel $L(z, \zeta)$, Henkin kernel $H(z, \zeta)$ with weight factors and a kernel $T(z, \zeta)$ and a differential form $P(z, \zeta)$ are constructed on Stein manifolds. Using the technique of localization, Qiu Chunhui [**Q**] proves that the principal value of integral with kernels B, L and H exist respectively. And then the main result is obtained, that is the kernels B, $L - B + T$ and $B + \bar{\partial}(f \wedge H)$ satisfy the fundamental equation

$$\bar{\partial} E = [\Delta] + P$$

respectively.

Secondly, Qiu Chunhui [Q] discussed the boundary behavior of the Bochner-Martinelli transform, Leray transform and Henkin transform with weight factors for arbitrary (p, q) differential form, and the fundamental solutions with weight factors of the induced Cauchy-Riemann equation (that is the $\bar{\partial}_b$-equation) were obtained.

REFERENCES

[AY]   Aizenberg, L. A. and Yuzhakov, A. P, *Integral representations and residues in multidimensional complex analysis*, Izdat. Nauk, New Siberia, 1979; English translation, Amer. Math. Soc., Providence, R.I., 1983.

[A]   Andersson, M., *Weighted solution formulas for the $\bar{\partial}$-equation on a Stein manifold*, preprint, Department of Mathematics, Chalmers University of Technology, The University of Göteborg, Sweden (1988).

[B]   Berndtson, B., *Cauchy-Leray forms and vector bundles*, preprint, Department of Mathematics, Chalmers University of Technology, The University of Göteborg, Sweden (1989).

[BA]   Berndtson, B. and Andersson, M., *Henkin-Ramirez formulas with weight factors*, Ann. Inst. Fourier (Grenoble) **32** (1983), 91–110.

[DL]   Demailly, J. P. and Laurent-Thiebaut, C., *Formules intégrales pour les formes différentilles de type (p, q) dan les variétés de Stein*, Ann. scient. Éc. Norm. Sup. **20** (1987), 579–598.

[HLa]   Harvey, F. R. and Lawson, H. B., *On boundaries of complex analytic varieties. I*, Ann. of Math. **102** (1975), 223–290.

[HP]   Harvey, F. R. and Polking, J., *Fundamental solutions in complex analysis. I: The Cauchy-Riemann operator*, Duke Math. J. **46** (1979), 253–300; II: The induced Cauchy-Riemann operator, Duke Math. J. **46** (1979), 301–340.

[HLe]   Henkin, G. M. and Leiterer, J, *Theory of Functions on Complex Manifolds*, Abteilung II, Bd. 60 (Mathematische Lehrbücher und Monographien), Akademie Verlag, Berlin, 1983.

[K]   Kakichev, V. A., *Character of the continuity of the boundary values of a Martinelli-Bochner integral*, Oblast. Ped. Inst. Uchen. Zap. 96 (Trudy Mat. Kafedr. 6) (1960), 145-160. (Russian)

[KR]   Kohn, J. J. and Rossi, H., *On the extension of holomorphic functions from the boundary of a complex manifold*, Ann. of Math. **81** (1965), 451–472.

[Lau]   Laurent-Thiebaut, C., *Formules intégrals et théorèmes du type "Bochner" sur une variété de Stein*, C. R. Acad. Sci. Paris (Série I) **295** (6 décembre 1982), 661–664.

[LinL] Lin Liangyu, *Boundary properties of Cauchy type integral on closed piecewise smooth manifolds*, Acta. Math. Sinica **31** (1988), 547–557. (Chinese)

[LinY] Lin Yaxian, *On the $\bar{\partial}_b$-equation on Stein manifolds*, Masters thesis, Xiamen University (1985). (Chinese)

[LZ]   Lu Qikeng and Zhong Tongde, *An extension of the Privalov theorem*, Acta. Math. Sinica **7** (1957), 144–165. (Chinese)

[Q]   Qiu Chunhui, *Fundamental solutions with weight factors of Cauchy-Riemann equations on Stein manifolds*, Master's thesis, Xiamen University (1987).

[RS]   Range, R. M. and Siu, Y. T., *Uniform estimates for the $\bar{\partial}$-equation on domains with piecewise smooth strictly pseudoconvex boundaries*, Math. Ann. **206** (1973), 325–354.

[S]   Sun Jiguang, *Singular integral equations on a closed smooth manifold*, Acta Math. Sinica **22** (1979), 675–692. (Chinese)

[WanX1] Wang Xiaoqin, *Fundamental solutions with weight factors of Cauchy-Riemann operator*, Master's thesis, Xiamen University (1985). (Chinese)

[WanX2] _____, *Singular integrals and analyticity theorems in several complex variables*, Doctoral dissertation, Uppsola University (1990).

[WanZ] Wang Zhiqiang, *Integral representation formulas for (p,q)-forms on real non-degenerate strictly pseudoconvex polyhedra*, Master's thesis, Xiamen University (1991). (Chinese)

[Wu] Wu Xiaoqin, *Integral representation of (p,q) differential forms with weight factors on complex manifolds*, Master's thesis, Xiamen University (1991). (Chinese)

[Y] Yao Zongyuan, *An integral formula on a bounded domain in $\mathbb{C}^n$*, Scientia Sinica (to appear). (Chinese)

[Z1] Zhong Tongde, *Boundary properties of integrals of Cauchy type for functions of several complex variables*, Acta Math. Sinica **15** (1965), 227–241. (Chinese)

[Z2] ———, *The singular integral equations on the smooth ori entable manifolds*, Journal of Xiamen University (Natural Science) **4** (1979), 1–8. (Chinese)

[Z3] ———, *Transformation formulae of singular integrals with Bochner-Martinelli kernel*, Acta Math. Sinica **23** (1980), 538–550. (Chinese)

[Z4] ———, *Regularization of the singular integral equation on characteristic manifold*, Kexue Tongbao **26** (1981), 1–6. (Chinese)

[Z5] ———, *Singular integral equations with Bochner-Martinelli kernel on smooth orientable manifolds*, Proceedings of the 1980 Beijing Symposium on Differential Geometry and Differential Equations, Science Press, Beijing, 1982, pp. 1709–1711.

[Z6] ———, *Integral Representations of Functions of Several Complex Variables and Multidimensional Singular Integral Equations*, Xiamen University Press, 1986. (Chinese)

[Z7] ———, *Holomorphic extension on Stein manifolds*, Report no. 10, Mittag-Leffler Institute (1987).

[Z8] ———, *On the extension of (p,q) differential forms from boundary,*, International Symposium of Number Theory and Analysis in Memory of Hua Lookeng, Vol. II, Springer-Verlag, 1991, pp. 337–343.

[Z9] ———, *Singular integral equations on Stein manifolds*, Journal of Xiamen University (Natural Science) (1991) (to appear).

[ZQ1] Zhong Tongde and Qiu Chunhui, *$\bar{\partial}$-closed extension of (p,q) differential forms in $\mathbb{C}^n$*, Journal of Xiamen University (Natural Science) (1991) (to appear).

[ZQ2] ———, *$\bar{\partial}$-closed extension of (p,q) differential forms on Stein manifolds* (to appear).

[ZH] Zhong Tongde and Huang Sha, *Complex analysis in several variables*, Hopei Educational Press, 1990.

DEPARTMENT OF MATHEMATICS, XIAMEN UNIVERSITY, XIAMEN, FUJIAN 361005, PEOPLE'S REPUBLIC OF CHINA

Recent Titles in This Series

(*Continued from the front of this publication*)

(See the AMS catalog for earlier titles)

234